RENEWABLE ENERGY
In the Global Context

About the Author

The author is currently a Director in the Ministry of New and Renewable Energy. During his professional career he has worked on different aspects of renewable energy programmes and policies, rural energy issues, Clean Development Mechanism under Kyoto Protocol to the UN Framework Convention on Climate Change, Global Environment Facility and bilateral and multilateral cooperation mechanisms. He provided support for technical cooperation among developing countries in the renewable energy sector and participated in the development and implementation of renewable energy programmes under Country Cooperation Framework. He joined The Energy and Resources Institute (TERI) in 2006 as Visiting Fellow for study on renewable energy. He holds a Ph.D. in Physics from Garhwal University, Uttarakhand. He has also co-authoured a book on Biogas and has contributed to many articles in Journals and at conferences. He lives in New Delhi with his wife and two children and also shares deep interest in theatre, philosophy and art.

RENEWABLE ENERGY
In the Global Context

P. C. Maithani

CONCEPT PUBLISHING COMPANY, NEW DELHI-110059

Cataloging in Publication Data—DK
Courtesy: D.K. Agencies (P) Ltd. <docinfo@dkagencies.com>

Maithani, P. C.
Renewable energy : in the global context / P.C. Maithani.
p. cm.
With special reference to India.
Includes index.
ISBN 13: 9788180695568
ISBN 10: 8180695565

1. Renewable energy sources—Government policy. 2. Renewable energy sources—Government policy—India. 3. Renewable energy sources Environmental aspects. 4. Renewable energy sources—Environmental aspects—India. I. Title.

DDC 621.042 22

ISBN-13: 978-81-8069-556-8
ISBN-10: 81-8069-556-5

First Published 2008

Published and Printed by
Ashok Kumar Mittal
Concept Publishing Company
A/15-16, Commercial Block, Mohan Garden
New Delhi-110059 (INDIA)

Phones : 25351460, 25351794
Fax : 091-11-25357103
Email : publishing@conceptpub.com

This book is dedicated to

My Parents,
Meena ji
&
My brother Rakesh

सचिव
भारत सरकार
नवीन और नवीकरणीय ऊर्जा मंत्रालय
SECRETARY
GOVERNMENT OF INDIA
MINISTRY OF NEW AND RENEWABLE ENERGY

सत्यमेव जयते

वी. सुब्रमणियन
V. Subramanian

Message

The expectations from renewable energy technologies have been growing worldwide as a solution for energy and global environment problems. Particularly, the climate change concerns have compelled the world community to work for the alternatives to the fossil fuel based energy systems and ultimately work towards zero carbon energy sources. Development of renewable energy technology (ies) is a step in this direction. However, renewables in their present form have limitations – both technical and commercial and would need significant developments before they surpass conventional fossil fuel based energy economy. In this context, concerted efforts are being made the world over, including in India, to make this energy transition happen in the shortest possible time frame. Specific to the Indian context, renewables have a pivotal role in reliable meeting the demand for energy services of all the sectors in a technically efficient, economically viable and environmentally sustainable manner.

2. India is generously endowed with renewable energy sources like solar, wind, biomass and small hydro resources. Low gestation periods for setting up renewable energy projects with quick returns, conducive government policies, continuous focus on research and development and availability of financing options are some of the key factors contributing to fast growth of the sector. In India, we have taken up ambitious plans for developing renewable energy. However, there exist many challenges and barriers for mainstreaming renewables. The "renewable revolution" supported by public incentives will in any case, depend on how rapidly the costs decline and efficiencies improve.

3. The book by Dr. P.C. Maithani is lucidly written so as to be understood even by an average reader without any background in technical and commercial knowledge. Present knowledge on the subject is based either on ignorance or misinformation. This book would help in placing the issues in the right perspective for proper awareness.

4. In his book, Dr. P.C. Maithani has outlined a strategy of research and development and policy initiatives, which would help in harnessing renewable energy in the country. It would also help in developing an understanding of the global renewable energy technologies and policy developments. I would consider this to be a valuable addition to the literature on energy.

(V. Subramanian)

Foreword

Reaching out commercial form of energy to the two billion poor people round the globe is the biggest challenge. To meet the energy demand of all households worldwide, energy supplies must be doubled by 2050. This is the most important finding of the World Energy Council's Energy Policy Scenarios to 2050. There is enough energy resource around the world to satisfy the energy demand over the next forty (+) years. The challenge will be to get these resources and energy services from where they are produced to the places that have the greatest need for them.

The second important finding is that while fossil fuels will continue to account for the largest proportion of primary energy requirements through the next four decades, we cannot only double world energy supplies and improve access but also effectively manage greenhouse gas emissions and address climate change.

The main driver for this change will be the higher energy prices. Higher prices will propel the developed world towards greater energy efficiency and attract much higher levels of capital investment in infrastructure. But massive new levels of public and private investment in research, development and deployment of clean and more efficient technologies are also urgently needed. This will also depend on the global carbon price that should be reasonably high to give equitable opportunity for the developed and developing countries.

The energy mix will become more diverse, provided government assumes a strong role in RD&D and the private sector cooperates. The world's energy mix will include more supplies of hydroelectricity, nuclear power with satisfactory planning for spent fuel, biofuels, biomass and other renewables.

Fossil fuels can remain a fixture in a low carbon economy, produced more efficiently with more effective management of greenhouse gas emissions. A new framework for agreement on setting a value for carbon is critical, since without strong international cooperation and government involvement greenhouse gas emissions will not be managed, stabilized and reduced.

Sustainable energy future can be attained through high level of global cooperation amongst industry and government, and deeper integration of regional and international energy markets. The major challenges of energy sustainability are: eradicating energy poverty, setting the global value for carbon, and establishing global rules of energy trading and investment amid growing energy nationalization. To attain this global energy community decided, at the 20th World Energy Congress, in November 2007, at Rome to focus future actions on:

(i) A global framework to curb greenhouse gas emissions beyond 2012 that will ensure a stable carbon price.
(ii) Global rules of energy trade and investment.
(iii) New financial schemes limiting investment risk and offering realistic returns.
(iv) More government engagement and public-private partnerships to address increasing global energy interdependence, a key strategy to eradicate poverty.

Whereas the energy leaders thought in November 2007 that greater reliance on fossil fuels will continue, the Bali Plan of Action in December 2007 reflects the resolve of governments round the globe to move towards low greenhouse gas scenario where local energy sources will play an increasing role. Though there is quite a similarity in approach but there is a finer difference in approach. If only the World Energy Congress had met after the Bali Conference, the resolve could have been more focused on renewables.

Eradicating energy poverty remains the main challenge for the global community. This has to be attained keeping in view the findings of the Fourth Assessment Report of the Intergovernmental Panel on Climate Change that the warming of the climate system is unequivocal, and that delay in reducing emissions significantly constrains opportunities to achieve lower stabilization levels and increases the risk of more severe climate change impacts. This effort will not in any way create barriers for the economic and social development and poverty eradication which are the global, regional and national priorities.

This paper is being written in the shadow of the celebrations on completion of 25 years of Government of India's focus on New and Renewable Energy Sources through the institutional arrangement of Commission of Additional Sources of Energy; Department of Non Conventional Energy Sources; and subsequently the Ministry of Non Conventional Energy Sources now renamed as the Ministry of New and Renewable Energy. Through these efforts not only India laid firm foundation for the renewable energy programme in the country, but also provided guidance to many emerging and smaller countries to establish their institutional arrangement for renewable energy promotion.

Work done in the country is not fully realised as the overall contribution of new renewables is small compared to other sources. Very few people would have realized the renewable energy based power generation capacity has contributed to 25 per cent of the total generation capacity added during the last five years (2002-07). This has happened in spite of the cost disadvantage but mainly through private entrepreneurs, duly supported by MNRE.

25 years is a long time in the history of any institution. It will be premature to make off-the-cuff remarks about functioning of the ministry. I am fortunate to have witnessed developments over the last 25 years and can appreciate efforts of the ministry. One can look back with pride the systematic growth in this sector in our country.

The Prime Minister of India decided to have the development of energy sector studied in an integrated manner. An Expert Committee on Integrated Energy Policy was set up under the

Planning Commission in 2005 that completed its report in August 2006. I had the privilege of being a member of the same.

The Committee realized that from a longer term perspective and keeping in view the need to maximally develop domestic supply options as well as the need to diversify energy sources, renewables remain important to India's energy sector. It is relevant to mention that solar power could be an important player in India attaining energy independence in the long run. One should not get lost in figures and comparative proportions of other energy sources. The Committee observed that with a concerted push and a 40-fold increase in their contribution to primary energy, the new renewables may account for only 5 to 6 per cent of India's energy mix in 2031-32. While this figure appears small, the distributed nature of renewables can provide many socio-economic benefits.

Renewables are often criticized for higher prices. In most of the cases users pay for the energy conversion equipment as well as the end use device. When government shares the investment through subsidy many classical economists become critical. Whereas they overlook the hidden subsidies in other forms of energy supplies but they cannot support subsidies to renewables. It is basically a matter of outlook of the person. The subsidies need to be linked to the service the renewables provide, and then the justification for subsidies can be easily appreciated.

Subsidies for renewables can be justified on several grounds. *A renewable energy source may be environmentally friendly.* Bali Action Plan (in December 2007) has called for enhanced action for the provision of financial resources and investment to support action on mitigation and adaptation and technology cooperation. It has specifically called for enhanced official and concessional funding for environment friendly and climate change mitigating and adapting technologies.

The second reason for justification for subsidy is the local availability thereby making it possible to supply energy earlier than in a centralized system. Grid connected renewables could improve the quality of supply and provide system benefits by generating energy at the ends of the grid where otherwise supply would have been lax.

Further renewables may provide employment and livelihood to the poor.

The Committee emphasized that the incentives should be linked to outcomes (energy generated) and not just outlays (capacity installed). Even when a capital subsidy is needed, it should be linked to outcomes. For example, capital subsidy could also be given in the form of a Tradeable Tax Rebate Certificate (TTRC) that could be based on actual energy generated. The rebate claim would become payable depending upon the amount of electricity/energy certified as having been actually supplied. MNRE deserves appreciation for announcing the scheme on similar lines. This scheme should streamline faster implementation of renewable electrification programme.

The Committee also emphasized on power regulators to create alternative incentive structures such as mandated feed-in laws or differential tariffs to encourage utilities to integrate wind, small hydro, cogeneration etc. into their system. The State Electricity Regulatory Commissions have not fully grasped the spirit of this recommendation. Decisions of commissions differ from state to state making implementation difficult. On the other hand the state governments consider local electricity generation as the localized activity and wish to leave it for the individual entrepreneurs to decide the purchase price. This attitude of SERCs has adversely affected expansion of renewable energy based electrification programme. A solution has to be found for this difficult situation.

Technologies for efficient use of traditional biomass have to be a social and economic agenda. Involvement of a number of ministries and agencies in this effort has not borne results. Intensive efforts are required.

Liquid bio fuels will hold the key for future development. Opposing this development is similar to opposition of India opening up its economy. The earlier we decide to enter the better it will be. One reason that our effort of utilizing wastelands for biomass production has not been successful is the unrealistic pricing that was projected from time to time. It should be understood that the wastelands can be utilized for biomass production by giving sufficient inputs and continuous nurturing. But, this will have the high first time cost. This costing needs to be carried out by commercial planters and professionals and not the research institutes. Otherwise the unrealistic pricing will deprive India of the opportunity to develop its own liquid bio fuels. Fertile lands producing cereal crops should not be touched otherwise that may result in food insecurity related issues. A word of caution: Either spend sufficient funds to develop wastelands or else the future will see bio fuel crops being grown on fertile lands that are now producing cereals. Economics will lure farmers to opt for bio fuel crop production.

A need was felt for the last 25 years to have an authentic and informative reference book that will give the global and national scenarios and growth patterns of renewable energy. Sporadic attempts were made. It is for the first time that this book presents a comprehensive picture of policies, strategies, methodologies and results of efforts in the renewable energy sector worldwide. I have yet to see any other similar effort.

Dr. P.C. Maithani, the author, has shown deep understanding of the subject and has presented the facts in a reader friendly manner, after researching original and authentic information from different countries and international agencies. I have no doubt readers will find the book a prized possession, and a valuable reference volume.

Dr. Maithani deserves all appreciation from the renewable energy fraternity.

Pradeep Chaturvedi
Chairman,
Indian Association for
the Advancement of Science

New Delhi

Preface

Steadily growing energy demand combined with local and global environmental concerns of climate change impacted by burning of fossil fuels, necessitates formulation of appropriate energy policies across the globe, including India. Renewable sources of energy are being focused to meet larger proportions of future energy needs. India has traditionally used renewable sources in one form or the other, so much so that the contribution of traditional biomass is about one third in India's primary energy supply. The Government realized the significance of renewable sources of energy and established a separate ministry to handle these sources.

In India introduction of renewable energy technologies has been considered one of the means to achieving sustainable energy transition. These sources have the potential to provide options to meet specific needs for energy services, particularly in rural areas and to create new employment opportunities. However, the question is what is the time frame within which renewable energy technologies would reach a level when these could be termed as safe, convenient, affordable and financially and economically viable? Obviously, at that stage no state intervention may perhaps be required and these technologies would compete with the existing baseline energy technologies. The task in hand is how to reach that stage in the shortest possible time frame. There are two possible routes: the first is think in competitive terms with conventional sources and penetrate at a slow pace, and the second is think innovatively and grow exponentially.

There are various projections of renewables programme in India. There is no doubt about the contribution of renewables in providing solutions for meeting the energy needs in the rural areas of the country, which could not have been thought of through any other means. Today, some of the renewable energy applications have started competing with fossil fuel based energy with little fiscal support. The potential of renewable energy in meeting the energy needs of the country can be substantial, but challenging. However, apprehensions have been expressed by policy makers, technology analysts and others about government support for renewables when cheaper energy options are available and also about the technological maturity of renewables. It is also said at many fora that why should India go for renewables when emissions reduction is not an issue for the country, as India has no emission reduction targets under the Kyoto Protocol to the United Nations Framework Convention on Climate Change. Although, India does not have any obligations, until now, for emission reduction, the Government of India, over the years, being a proactive player. It has legislated and mandated appropriate policies to promote renewables.

In shaping the contours of the Indian government's national renewable energy development programme experts, NGOs, multilateral banks and international organizations need to provide scenarios and options after understanding the local socio-economic fabric. This has to be backed by world class R&D and a deployment programme that is responsive to different scenarios. However, inadequate domestic research effort in combination with a rather limited debate among major domestic stakeholders and strong market interests leaves the contours of the Indian renewable energy to be, by and large, grid-based renewable power centric.

This book seeks to develop an effective understanding of the renewable energy technologies and policy developments to foster an era of renewable energy, especially in India in an integrated manner to meet the energy and environmental challenges facing the nation. These challenges include the gap between demand and supply of electric power, non-availability of modern forms of energy to a majority of rural population, global climate change, air pollution and oil dependence.

The book has been organized in seven chapters as under:

Chapter I: Introduction: This chapter provides a primer to present global energy scene, future energy demand scenarios and development of energy policy.

Chapter II: Renewable Energy Technologies: Present Status and Future Projections: This provides an overview of the present status and future projections for major renewable energy technologies—solar photovoltaic, solar thermal power, wind electric, biomass power, bio-fuel and hydrogen in terms of technological development and cost reduction.

Chapter III: Renewable Energy Policies in Other Countries: Analysis and review of existing renewable energy policies of EU, Germany, USA, Japan and China have been made, which are implementing fairly large renewable energy programme.

Chapter IV: Renewable Energy in India: This chapter analyses the past developments, present policies of Indian renewable energy status and programmes. It explores various stages of renewable energy development in the country and evaluation of policies and programmes.

Chapter V: Renewable Energy and Environment: This chapter discusses the environmental externalities and cost arising due to impacts of energy generation, transmission and consumption and quantification of such cost that could be accounted for in determining cost of energy. Adding such costs to direct energy costs, as being done by many of the national governments, could make renewable energy a competitive option. Analysis of Indian renewable energy projects under the carbon credit mechanism has also been presented.

Chapter VI: Strategy for Research and Development: This chapter provides an analysis of research and development efforts in renewable energy and develops a case for an alternative technology development strategy for renewable energy within the country.

Chapter VII: Renewable Energy Policy Directions: This

chapter provides policy suggestions which would help in building a long-term renewable energy portfolio.

This is my sincere effort to describe global actions and their adaptation in Indian context, thereby making an important complement to the literature on renewable energy technologies and policy. It refers to works of individuals and organizations and that have been referenced. The book is a quick reference for renewable energy professionals, investors and businesses and policy makers.

P.C. Maithani

Acknowledgements

As a member of the Indian delegation to the Eighth Conference of Parties to the United Nations Framework Convention on Climate Change, I was moved and deeply touched by the expectations of the delegates from all around the world on renewables as a real solution for addressing climate change issues. This inspired me to plan for a sabbatical from the rigmarole of handling government work and undertake a study that collates the available work on present renewable technological status, future projections and representative policy regimes at one place, examine India's position and put renewables in its rightful perspective instead of their claiming the sky. By that time I had already worked for a decade and half on various aspects of renewable energy technology, its development and dissemination, in close association with a number of professionals and institutions across the world. I sincerely convey my gratitude to all of them and hope that the insight to the issues, learning I had from these associations has been reflected, at least in fraction, in this study.

I would like to express my gratitude to Dr. R.K. Pachauri, Director General, The Energy and Resources Institute (TERI), who gave me an opportunity to pursue my research and study in TERI's stimulating environment during my sabbatical. His guidance as a teacher whenever I visited him, often unscheduled, and at the spur of the moment has given me inspirational insights into many of the issues under study. I often felt that I was stealing into his time. I would also thank to my distinguished colleagues in TERI including Air Cmde M.M.Joshi, Col. R.P. Singh, Shri Amit Kumar, Dr. V.V.N. Kishore, Dr. Sameer Maithel, Dr. Y.P. Abbi, Shri Mahesh Vipradas, Dr. J.S. Rawat, Ms Akanksha Chaurey and Ms Parimita Mohanty for their valuable help and support. Thanks to Shri Gokul Ramamurthy, Ms Isabelle Noirot and Dr. Haruko Ishikawa who were very kind to me for providing continued support. I am also thankful to Ms Rita who gave me typing support whenever I requested for the same.

I would like to place on record my sincere gratitude to the Ministry of New and Renewable Energy (MNRE) with whom I am employed. MNRE has never remained a conventional bureaucratic government Ministry, it rather is a hub where my seniors and colleagues always found pleasure in developing programmes and policies that leads to renewable energy technology development and harnessing its potential for various end use applications. Although, at times, we had our own share of frustrations, but generally the work environment is very informal filled with enthusiasm and dreams. With out my being in MNRE, it would have been impossible for me even to think of undertaking such a task.

I would like to thank Shri V. Subramanian, Secretary, MNRE, who even with his busy schedule took time to read the manuscript and provided me with valuable suggestions. My sincere gratitude to Shri Ajit K. Gupta, Adviser, MNRE who not only twice read the manuscript but also spent substantial time in suggesting me revisions. I would like to thank Shri D. Majumdar, Chairman and Managing Director, IREDA for his suggestions. I am also thankful to Dr. Ahmar Raza, Director and Shri Gangesh Upadhadhay, Principal Scientific Officer for their always-available help and support.

I am grateful to Dr K.C. Khandelwal, former Adviser, with whom I started my career in the Ministry, for his continued guidance and affection. I am thankful to Shri A.K. Mangotra, former Joint Secretary, MNRE for his suggestions on the study, continued interest in my work and affection. I am also thankful to Shri Sunil Khatri, former Joint Secretary, MNRE for providing me insight of system analysis and also shaping up the initial study proposal. I also thank Smt. Shakuntala D. Gamlin, former Director, MNRE for her continued support all along.

I am grateful to Dr. Venkata, World Bank; Dr. Pallav Purohit, IIASA; Prof. Ashok Parthsarathi, JNU; Dr. A.P. Kulshrestha, NAM S&T Centre; Prof. Jyoti Parikh, IRADe; Dr. Ajay Mathur, BEE; Dr. S.K. Panigrahi, Planning Commission; Ms Pamposh Bhat, GTZ and Dr. Bhaskar Natarajan, ICEF for their interest in my work, help and suggestions.

I would like to thank Shri Chittoor P. Krishnan, a person with immense knowledge and wisdom, for his help in editing of the manuscript and also contributing in many other ways.

My friend Dr. Arjun Narayan and elders Shri Chander Mohan, and Shri B.K. Bhatt have been unconditional in their love and emotional support throughout. Chander, a great professional has shown wisdom far above of his age, at those times when my perspectives during the course of study went astray.

Thanks are due to Shri Pradeep Chaturvedi, for his interest in my work and un-bounding affection. Many discussions I have had with him helped me to gain useful insights. He also went through the manuscript and made many suggestions.

I am grateful to the Shri A.K. Mittal and Shri Nittin Mittal of Concept Publishing Company for their personal interest in accepting to publish this book and reach it to the wider public at large.

This work would not have been possible without the support of my wife Vijaya, daughter Prachi and son Patanjali. They have suffered my long absences from home without any complaint. I often disturbed Prachi asking her to correct my English. They always withstood me and continued to be my strength. Finally, my parents and siblings have eternally been the soul of my internal energy, drive and passion. I will continue to draw my strength from them.

Contents

List of Acronyms

ACT	—	Accelerated Technology
ALCC	—	Annualized Life Cycle Cost
a-Si	—	Amorphous Silicon
BHEL	—	Bharat Heavy Electricals Limited
BIS	—	Bureau of Indian Standards
Bkwh	—	Billion Kilowatthours
bm^3	—	Billion Cubic Meters
BMU	—	Federal Environment Ministry
BMWi	—	Ministry for Economy and Technology
BPL	—	Below Poverty Line
CASE	—	Commission for Additional Sources of Energy
CCS	—	CO_2 capture and storage
CDM	—	Clean Development Mechanism
CdTe	—	Cadmium Telluride
CEFV	—	Clean Energy Financing Vehicle
CHP	—	Combined Heat and Power
CM	—	Combined margin
CO_2	—	Carbon-dioxide
CPUC	—	California Public Utilities Commission
CSD	—	Commission on Sustainable Development
CSI	—	California Solar Initiative
CSP	—	Concentrating Solar Power
C-WET	—	Centre for Wind Energy Technology
DNES	—	Department of Non-Conventional Energy Sources
DoE	—	Department of Energy
DSIR	—	Department of Scientific and Industrial Research
EC	—	European Commission

EU	—	European Union
ECT	—	Energy Charter Treaty
EEG	—	Erneuerbare-Energien-Gesetz
EFL	—	Electricity Feed Law
EIA	—	Energy Information Administration
EPACT	—	Energy Policy Act
ETA	—	Energy Tax Act
FAO	—	Food and Agriculture Organization
FBR	—	Fast Breeder Reactor
FERC	—	Federal Energy Regulatory Commission
FFA	—	Free Fatty Acid
FOB	—	Free-On-Board
FT	—	Fischer-Tropsch
GATT	—	General Agreement on Trade and Tariff
GDP	—	Gross Domestic Product
GEF	—	Global Environment Facility
GERC	—	Gujarat Electricity Regulatory Commission
GHG	—	Green House Gases
GREET	—	Global Renewable Energy Education and Training
GROWIAN	—	Große Windenergieanlage Bau-und-Betriebsgesellschaft mbH
GW	—	Gigawatts
IARI	—	Indian Agricultural Research Institute
IEA	—	International Energy Agency
IIASA	—	International Institute for Applied Systems Analysis
IOC	—	Indian Oil Corporation
IPCC	—	Intergovernmental Panel on Climate Change
IPHE	—	International Partnership on Hydrogen Economy
IREDA	—	Indian Renewable Energy Development Agency
JPOI	—	Johannesburg Plan of Implementation
KERC	—	Karnataka Electricity Regulatory Commission
KVIC	—	Khadi and Village Industries Commission

kW	—	Kilowatts
LCA	—	Life Cycle Analysis
LEC	—	Levelized Electricity Cost
MDGs	—	Millennium Development Goals
MERC	—	Maharastra Electricity Regulatory Commission
METI	—	Ministry of Economy, Trade and Industry
MITI	—	Ministry of International Trade and Industry
MNRE	—	Ministry of New and Renewable Energy
MOF	—	Ministry of Finance
MOST	—	Ministry of Science and Technology
MPERC	—	Madhya Pradesh Electricity Regulatory Commission
MSW	—	Municipal Solid Waste
mtoe	—	Million Tones of Oil Equivalent
NAL	—	National Aeronautical Laboratory
NCRPV	—	National Coordinated Research on Solar PV
NCST	—	National Committee on Science & Technology
NDRC	—	National Development and Reform Commission
NEA	—	National Energy Act of 1978
NEDO	—	New Energy Development Organization
NGOs	—	Non-Governmental Organisations
NGRI	—	National Geophysical Research Institute
NHEP	—	National Hydrogen Energy Roadmap
NOVOD	—	National Oilseeds and Vegetable Oils Development
NPBD	—	National Project on Biogas Development
NPL	—	National Physical Laboratory
NRSA	—	National Remote Sensing Agency
NRSE	—	New and Renewable Sources of Energy
NSS	—	National Sample Survey
OECD	—	Organisation for Economic Co-operation and Development
OERC	—	Orissa Electricity Regulatory Commission

OM	—	Operating Margin
OTEC	—	Ocean thermal energy conversion
PEM	—	Proton Exchange Membrane
PHWR	—	Pressurized Heavy Water Reactors
PIER	—	Public Interest Energy Research
PPP	—	Purchasing Power Parity
PTC	—	Production Tax Credit
PURPA	—	Public Utility Regulatory Policies Act of 1978
PV	—	Photovoltaic
R&D	—	Research and Development
RD&D	—	Research, Design and Development
REDB	—	Rural Electrification Distribution Backbone
REN21	—	Renewable Energy Policy Network for 21st Century
RERC	—	Rajasthan Electricity Regulatory Commission
RES	—	Renewable Energy Sources
RES-E	—	Renewable Energy Sources
RGGVY	—	Rajiv Gandhi Grameen Vidyutikaran Yojana
RPS	—	Renewable Portfolio Standards
SACC	—	Scientific Advisory Committee of the Cabinet
SAI	—	Solar America Initiative
SAR	—	Second Assessment Report
SEBs	—	State Electricity Boards
SERC	—	State Electricity Regulatory Commissions
SHP	—	Small Hydro Power
SONATA	—	Son-Narmada-Tapi
SRES	—	Special Report on Emissions Scenarios
SVOs	—	Straight Vegetable Oils
T&D	—	Transmission and Distribution
TBOs	—	Tree Based Oilseeds
TCES	—	Total Commercial Energy Supply
TDB	—	Technology Development Board
TIFAC	—	Technology Information, Forecasting and Assessment Council
TNERC	—	Tamil Nadu Electricity Regulatory Commission

TPES	—	Total Primary Energy Supply
TWh	—	Terawatthours
UCE	—	Unit cost of electricity generation
UERC	—	Uttarakhand Electricity Regulatory Commission
UNCED	—	United Nations Conference on Environment and Development
UNDP	—	United National Development Programme
UNESCO	—	UN Educational, Scientific and Cultural Organization
UNFCCC	—	Nations Framework Convention on Climate Change
UNFCCC	—	United Nations Framework Convention on Climate Change
UNFPA	—	United Nations Population Fund
UNIDO	—	United Nations Industrial Development Organization
UPERC	—	Uttar Pradesh Electricity Regulatory Commission
WEC	—	World Energy Council
WHO	—	World Health Organization
WMO	—	World Meteorological Organization
WSSD	—	World Summit on Sustainable Development
WSSP	—	World Solar Summit Process
WTO	—	World Trade Organization

List of Figures and Tables

Figures

Tables

Chapter 1

Introduction

Energy creates the foundation that supports a nation, from the economy and industry to everyday livelihood, and significantly affects the issue of national security. Energy use and supply are of fundamental importance to society and it is considered an essential precondition for human development. From the earliest times when wood was used for light and heat several of centuries ago by primitive man, to the latest energy technologies, greater quality and efficiency in energy use has been the aim of and driving force behind innovation and progress. The first shift was use of coal-fired steam engines, which allowed for new, more efficient production processes and simultaneously reduced dependence on traditional biomass fuels. The second shift was from coal to oil and gas that increased mobility through introduction of combustion engine for transport and other purposes. In parallel and the third, was use of energy carrier in the form of electricity that made possible the information age through widespread application of computers and internet. Within the last century the use of solid and liquid fossil fuel energy grew several folds. However, there still exist wide gaps in terms of energy use pattern among different parts of the world, with one third of the global population still relying on traditional biomass in its basic form for meeting their energy needs.

With the beginning of industrialization in early 1800, there has been significant a change in global development in terms of factors of growth and energy use. Some estimates indicate that during the last 200 years global energy use has increased by 35 times and energy related carbon emissions have increased by a factor of 22 (Table 1.1).[1]

Table 1.1 Factors of growth—The Last 200 Years

	Year		*Factor (x times)*
	1800	*2000*	
World Population (billion)	1	6	6
Life Expectancy, Years	35	75	2
Work hours per year	3000	1500	½
World Income, Trillion $	0.5	36	72
Global Energy Use, Gtoe	0.3	10	35
Carbon, Energy, GtC	0.3	6	22
Carbon all sources, GtC	0.8	8	10

The global energy mix has also witnessed major shifts as mentioned earlier. In last 100 years, there were drastic changes in the sources of energy (Figure 1.1). At the same time, due to large scale and pervasive nature of energy related activities, energy use has made the greatest impact on the environment of any human activity. Whereas in the initial stages energy and environment concerns were perceived of their local in nature relating to extraction, transport or noxious emissions, their impacts have now taken a global dimension covering issues such as acid rains and the greenhouse effect. These problems have drawn the attention of the political and scientific community the world over and have become a subject of intense international debate and regulation.

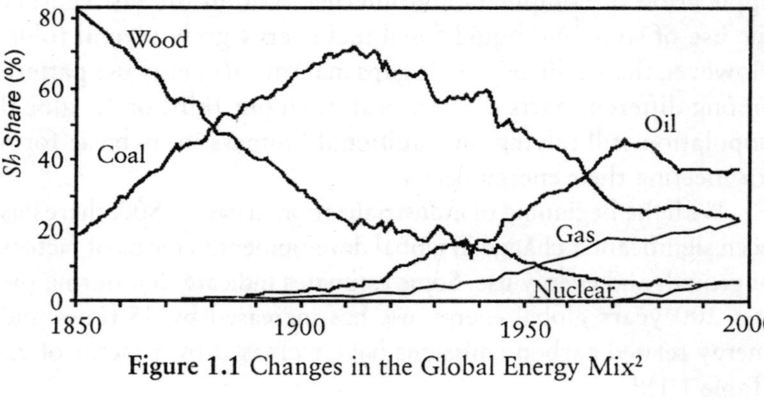

Figure 1.1 Changes in the Global Energy Mix[2]

1.1 Present Global Energy Scene

The total primary energy supply (TPES) has grown by over 86 per cent from 1973 to 2005. The share of fossil fuel all along has remained predominant (Table 1.2). Over these years, the TPES of Organisation for Economic Co-operation and Development (OECD)[3] countries continued to increase in absolute terms; however, its share in overall TPES has decreased from 61.3 per cent to 48.5 per cent. (Table 1.3) Major portion of renewables and wastes combustible (traditional biomass) continued to be used in non-OECD countries. Commercial non-hydro renewable energy makes very small contribution in the TPES.[4]

Table1.2 Evolution of World Total Primary Energy Supply by Fuel (Mtoe)

Fuel	*1973 (%)*	*2005 (%)*
Coal	24.4	25.3
Oil	46.2	35.0
Natural Gas	16.0	20.7
Nuclear	0.9	6.3
Hydro	1.8	2.2
Combustible Renewable and Waste	10.6	10.0
Others (geothermal, solar, wind etc.)	0.1	0.5
Total	6128 Mtoe	11435 Mtoe

Table 1.3 World Total Primary Energy Supply by Region

Region	*1973 (%)*	*2005 (%)*
OECD	61.3	48.5
Former USSR	14.2	8.6
China	7.1	15.2
Asia (excluding China)	5.6	11.2
Africa	3.5	5.3
Other	6.2	11.2
Total	6128 Mtoe	11435 Mtoe

There has been significant growth in electricity generation during the period. The total electricity generation has grown by around 198 per cent from 1973 to 2005 (Table 1.4). The share of fossil fuel all along has remained predominant. Over these years, the electricity generation of OECD countries continued to increase in absolute terms, however its share in overall electricity has decreased from 72.9 per cent to 56.9 per cent (Table 1.5). The share of non-hydro renewable electricity increased over 3 times from very modest 0.6 per cent to 2.2 per cent.

Table 1.4 World Electricity Generation by Fuel (TWh)

Fuel	*1973 (%)*	*2005 (%)*
Coal	38.3	40.3
Oil	24.7	6.6
Gas	12.1	19.7
Nuclear	3.3	15.2
Hydro	21.0	16.0
Others (geothermal, solar wind, biomass etc)	0.6	2.2
Total	6116 TWh	18235 TWh

Table 1.5 Regional Shares of Electricity Generation

Region	*1973 (%)*	*2005 (%)*
OECD	72.9	56.9
Former USSR	15.0	7.7
China	2.8	13.9
Asia (excluding China)	2.6	8.8
Africa	1.8	3.1
Other	4.9	9.6
Total	6116 TWh	18235 TWh

As about four-fifths of green house gases (GHG) emissions are energy related, energy generation and consumption are key determinants of carbon-dioxide (CO_2) emissions. The CO_2 emissions from energy related activities have grown by around 73 per cent from 1973 to 2005 (Table 1.6). The CO_2 intensity has not improved in real terms as it remains 2.37 Mt CO_2/Mtoe

TPES in 2005 as against 2.59 Mt CO_2/Mtoe in 1973. Over these years, the CO_2 emissions from OECD countries continued to increase in absolute terms, however its share in overall emissions has decreased from 65.9 per cent to 47.6 per cent (Table 1.7).

Table 1.6 World CO_2 Emissions by Fuel (Mt of CO_2)

Fuel	*1973 (%)*	*2005 (%)*
Coal	34.9	40.5
Oil	50.7	39.5
Gas	14.4	19.7
Others (industrial wastes etc)	—	0.3
Total	15661 Mt	27136 Mt

Table 1.7 Regional Shares of World CO_2 Emissions

Region	*1973 (%)*	*2004 (%)*
OECD	65.9	47.6
Former USSR	14.4	8.5
China	5.7	18.8
Asia (excluding China)	3.0	9.5
Africa	1.9	3.1
Other (including bunkers)	9.1	12.5
Total	15661 Mt	27136 Mt

1.2 Future Energy Demand Projections

Many institutions and organizations have developed numerous long-term energy scenarios which, invariably, are based on a definite story line and assumptions on key variables such as the global population, economic growth, technology development and diffusion, global and regional economic integration, fuel prices, agriculture, land-use patterns, etc. In order to have a relative comparison here three most cited scenarios have been narrated.

1.2.1 IIASA-WEC Study

International Institute for Applied Systems Analysis (IIASA) and the World Energy Council (WEC) study was undertaken during the period 1993-98.[5] In this study three main scenarios with principal focus on the period up to 2050 were presented and projections were also made out to 2100. In brief, Case A presents a future of impressive technological improvements and consequent high economic growth. Case B describes a future with less ambitious, technological improvements and consequently with intermediate economic growth. Case C presents an ecologically driven future. It includes both substantial technological progress and unprecedented international cooperation centered explicitly on environmental protection and international equity.

Across the three cases, the primary energy needs increase up to threefold by 2050 and up to fivefold by 2100. Despite the divergence in primary energy structures, the pattern of final energy use is predicted to be consistent across scenarios, showing a continuing trend toward energy reaching the consumer in ever more flexible, more convenient, and cleaner forms. World final energy by form, in per cent, as solids, liquids, and grids is indicated in Figure 1.2 (Overlapping shaded areas indicate variations across Cases A, B, and C).

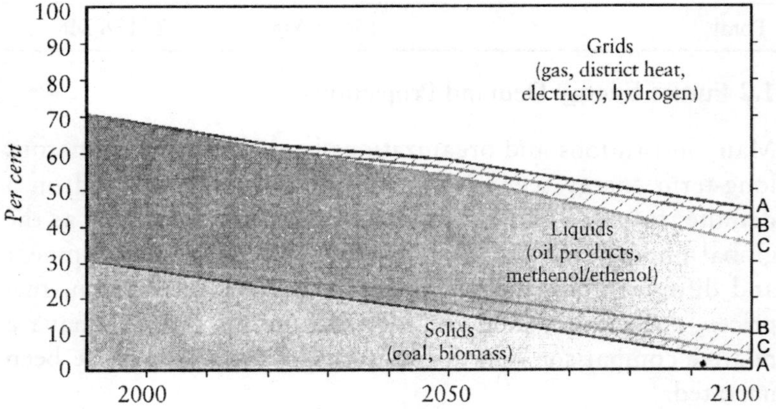

Figure 1.2 IIASA-WEC World Energy Mix Projections

1.2.2 Special Report on Emissions Scenarios 2000 and IPCC-SAR

In the Intergovernmental Panel on Climate Change (IPCC)'s

Special Report on Emissions Scenarios (SRES), projected world primary energy use in 2100 fell within a range of 600 to 2800 EJ for 90 per cent of the scenarios explored. Among the scenarios surveyed in SRES, the average annual growth rates for global energy demand over the period from 2000 to 2100 ranged from 2.4 per cent per year to –0.1 per cent per year, with a median value of 1.3 per cent per year. The renewable energy share (including large hydro) in global primary energy is projected to be between 8 per cent and 17 per cent under alternate scenarios in 2030. The share increases to above 61 per cent under the one of the scenarios in 2100, with biomass supplying 17 per cent and resulting in almost 4 per cent of global land use devoted to biomass for energy purposes. In 2030, the expected demand increase is relatively similar in all scenarios, and while the share of renewable energy differs marginally the share in 2030 remains relatively small in all cases (Figure 1.3).

In IPCC-SAR (Second Assessment Report) a review of medium-term (to 2025) and long-term (up to 2100) potentials of renewable energy has been made (Table 1.8)[6].

Table 1.8 Global Renewable Energy Potentials for 2020 to 2025, Maximum Technical Potentials, and Annual Flows, in EJ

	Consumption		*Potentials by*	*Long-term technical potentials*	*Annual flows*
	1860-1990	1990	2020-2025		
Hydro	560	21	35-55	> 130	> 400
Geothermal	—	—	4	> 20	> 800
Wind	—	—	7-10	> 130	> 200,000
Ocean	—	—	2	> 20	> 300
Solar	—	—	16-22	> 2,600	> 3,000,000
Biomass	1,150	55	72-137	> 1,300	> 3,000
Total	1,710	76	130-230	> 4,200	> 3,000,000

Figure 1.3 SRES—World Primary Energy Projections (EJ) Under Alternate Scenarios

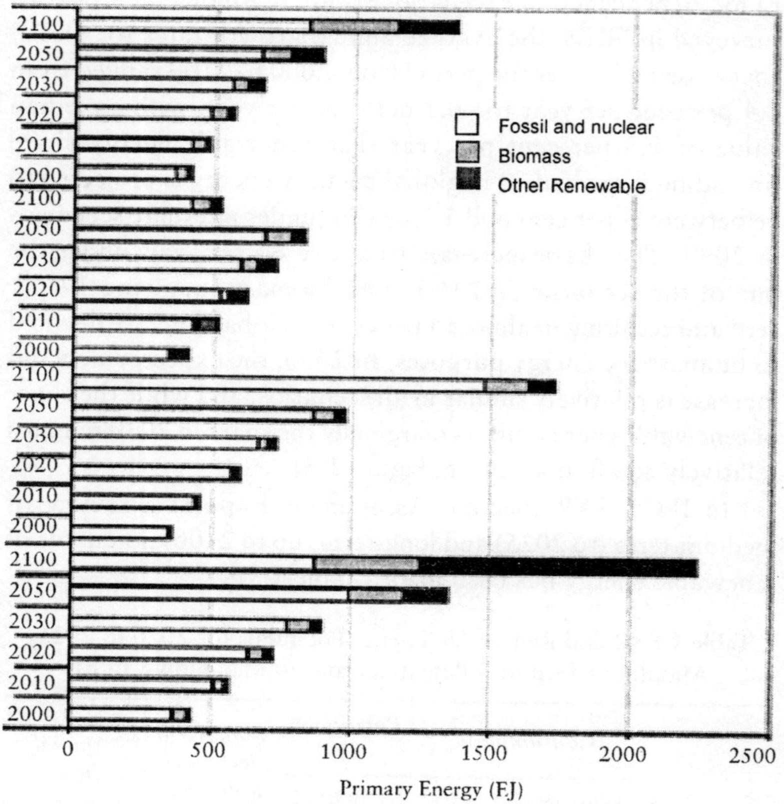

1.2.3 International Energy Agency

In 2005, IEA initiated a work to map the energy demand and likely fuel scenario for 2050 and the results of the study have been summed up in a book *Energy Technologies Perspective: Scenarios and Strategies 2050.*[7] In this study, apart from reference scenario, different alternate cases, which are called Accelerated Technology scenarios (ACT—for short)[8], have been developed to see the impact of various measures. The technology areas where different assumptions are made to model speed of energy efficiency gains and technology market penetration are: (1) the progress in cost reductions for renewable power generation technologies; (2) constraints on the development of nuclear power plants; (3) the risk that CO_2 capture and storage (CCS) technologies will not be commercialized by 2050; and (4) the effectiveness of policies to increase the adoption of energy efficient end-use technologies. In addition, a 'TECH Plus' scenario, has been analysed in which more significant technical progress and cost reductions are assumed for certain technologies. It is the latest and most elaborate study which analyses alternative scenarios taking energy use as per present trends.

1.2.3.1 Energy Demand

In the Baseline Scenario, total primary energy supply (TPES) grows at 1.6 per cent on average per year, from 10579 Mtoe in 2003 to 22112 Mtoe in 2050. This rate of growth is less than the 2.1 per cent per year that occurred between 1971 and 2003, but it still represents an increase of 109 per cent in primary energy demand between 2003 and 2050. Between 2003 and 2050 the production of coal triples, that of natural gas more than doubles and that of oil almost doubles (including synfuels). By 2050, coal becomes the predominant fuel and accounts for 34 per cent of primary energy use in 2050. It surpasses oil demand in absolute terms between 2030 and 2050. The share of oil in TPES declines from 34 per cent in 2003 to 27 per cent in 2050. Natural gas increases its share from 21 per cent in 2003 to 24 per cent in 2050. Non-fossil fuels account for just 15 per cent of demand in 2050, down from 20 per cent in 2003. Of the non-fossil fuels, nuclear energy share declines from 6 per cent in 2003 to 4 per cent in 2050, other renewables from 11 per cent in 2003 to 9 per cent, while hydro remains at 2 percent. The share of fossil fuel in the total demand increases from 80 per cent in 2003 to 85 per cent in 2050, despite the growth in nuclear and renewable energy. It follows that concerns about energy security will continue, as well as concerns associated with the continued growth in CO_2 emissions. In the Baseline Scenario, the share of developing countries in global energy demand rises considerably, from 38 per cent in 2003 to 54 per cent in 2050. OECD countries' share of primary energy use falls from just over half in 2003 to 38 per cent in 2050. In regard to the regional distribution in the baseline scenario, around 55 per cent of primary energy will be used in developing countries.

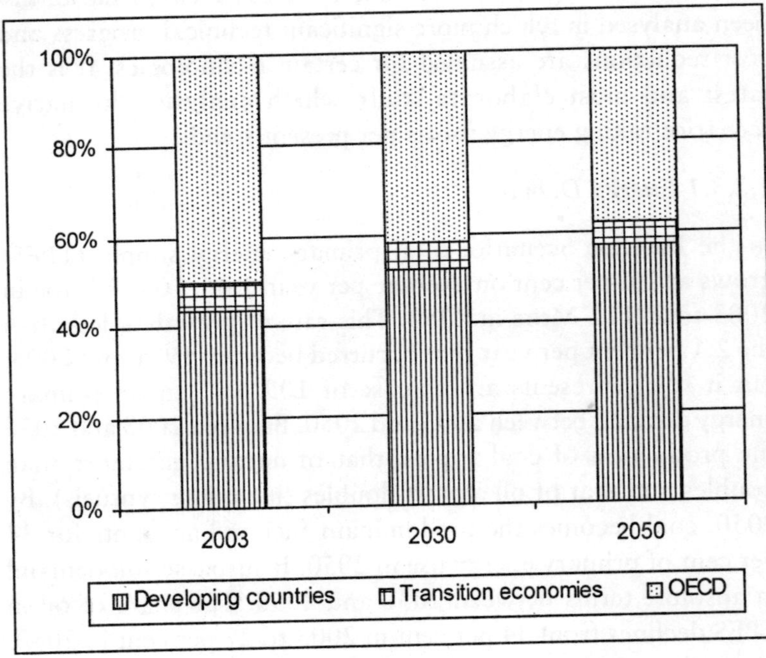

Figure 1.4 Primary Energy use Share by Region in the Baseline Scenario

In the Map scenario[9], primary energy use in 2050 is 58 per cent higher than in 2003, at 16762 Mtoe. This is 24 per cent lower than that in the Baseline Scenario in 2050. The use of fossil fuels in 2050 is significantly lower in all the ACT scenarios than in the Baseline Scenario, ranging from 28 per cent lower in the Low Efficiency scenario to 42 per cent lower in the no CCS scenario. Primary oil supply in the ACT scenarios is from 7 per cent to 21 per cent lower and gas demand is between 16 per cent and 30 per cent lower in 2050.

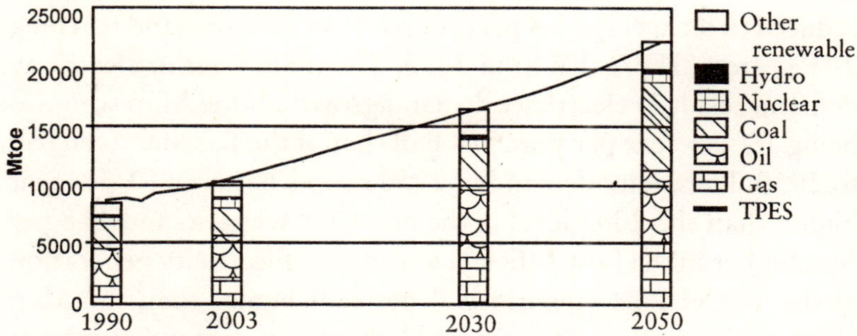

Figure 1.5 World total Primary Energy Supply by Fuel in the Baseline Scenario

1.2.3.2 Electricity

Electricity demand growth in the Baseline Scenario is on average 2.2 per cent per year between 2003 and 2050, making electricity the fastest growing component in total final demand. Electricity demand increases from 1433 Mtoe (16661 TWh) in 2003 to 4010 Mtoe (46631 TWh) in 2050. Electricity's share of final demand increases from 16 per cent in 2003 to 23 per cent in 2050. These trends are driven by rapid growth in population and incomes in developing countries, by the continuing increase in the number of electricity consuming devices used in homes and commercial buildings, and by the growth in electrically driven industrial processes in industry.

In the Baseline Scenario, coal-based generation in 2050 is more than three-times higher than present level and accounts for more than half of the increased production between 2003 and 2050. The capacity of gas-fired power plants quadruples and they account for about a third of increased global production. Hydro and non-biomass renewables (mostly wind) contribute about 6 per cent each to the new capacity required, while biomass and nuclear each account for around 2 per cent of the increased capacity.

Due to the higher population growth and the rapid increases in GDP and per capita incomes, baseline electricity demand in developing countries grows on average 3.6 per cent per year, three times faster than in OECD countries.

In the Map scenario, global electricity demand growth is reduced to on average 1.4 per cent per year, with demand reaching 2732 Mtoe (31776 TWh) in 2050. The reductions in electricity demand result in electricity demand growth in the Map scenario being 1.1 per cent per year, just half that of the Baseline Scenario to 2050. Electricity demand in 2050 ranges between 81 per cent higher than the 2003 level in the No. CCS scenario and 118 per cent higher in the Low Efficiency scenario. Electricity generation in the TECH Plus scenario is 4 per cent higher than the Map scenario in 2050 due to increased electricity demand to power decentralized electrolysers to produce hydrogen.

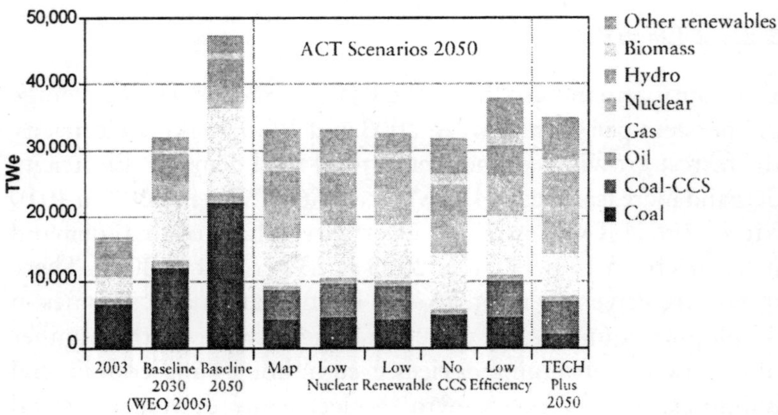

Figure 1.6 Global electricity production by fuel and scenario, 2003, 2030 and 2050

1.2.3.3 Electricity Generation by Region

In the Map scenario, non-OECD regions account for over 90 per cent of the increase in electricity generation between 2003 and 2050, with generation reaching 20752 TWh in 2050, which is about twice the OECD level. Almost a third of generation in non-OECD regions is coal-based, of which 45 per cent is from plants equipped with CCS. Another third is from renewables of which hydro contributes almost half. The remainder is from gas (24 per cent) and nuclear (11 per cent). In OECD regions total coal-fired generation in the Map scenario declines by 1580 TWh between 2003 and 2050, and coal only constitutes 21 per cent of total generation in 2050, of which plants with CCS provide three-quarters. Gas-fired generation in OECD increases 560 TWh from 2003 levels and reaches the same level as coal by 2050. Nuclear contributes about 28 per cent of OECD generation, hydro accounts for 15 per cent and non-hydro renewables for 13 per cent.

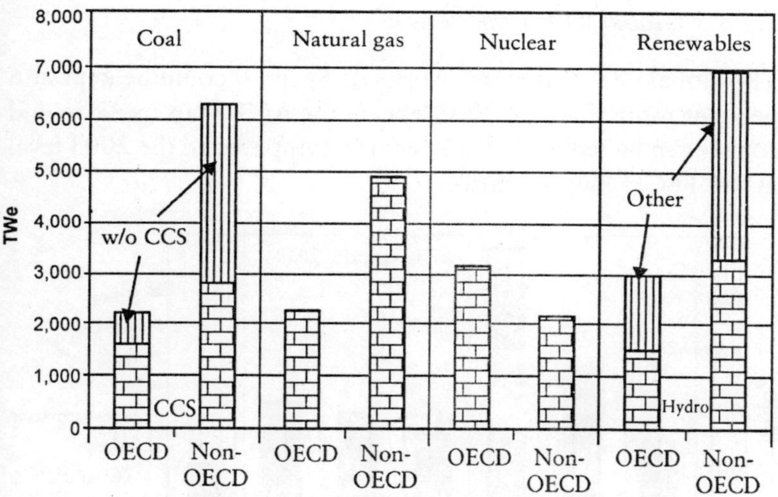

Figure 1.7 Electricity generation by fuel in the Map scenario in OECD and non-OECD regions, 2050

1.2.3.4 Electricity Generation from Renewables

By 2050, the increased use of renewables such as hydropower, wind, solar and biomass in power generation contributes between 9 per cent and 16 per cent of the CO_2 emission reductions in the ACT scenarios. The share of renewables in the generation mix increases from 18 per cent today, to as high as 34 per cent by 2050. In a scenario with less optimistic assumptions about cost reductions for renewable technologies, their share of generation is 23 per cent in 2050. On the other hand, in the TECH Plus scenario, which is more optimistic for both renewable and nuclear technologies, the share of renewables reaches more than 35 per cent by 2050. In most of the scenarios, wind is second to hydropower as the most important renewable source. Solar photovoltaic (PV) technology is playing a rapidly growing role in niche applications. By 2050, however, solar energy (PV and CSP) share in global power generation will still be below 2 per cent in all the ACT scenarios.

1.2.3.5 Global CO_2 Emissions

The global CO_2 emissions (Mt CO_2) by 2050 could be kept at 6 per cent more than the 2003 level in the ACT Map scenario and could even be reduced by 16 per cent compared to the 2003 level in the TECH Plus scenario.

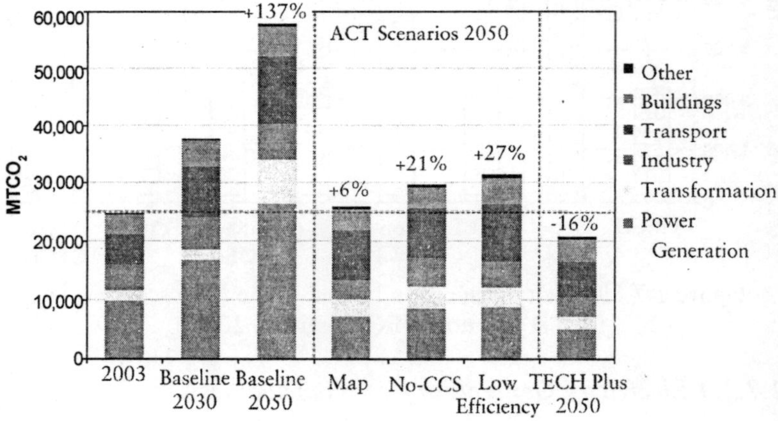

Figure 1.8 Global CO_2 Emissions 2003-2050 Baseline, ACT and TECH plus Scenarios

1.3 Towards Global Energy Policy

For a long period energy policy has been mainly the responsibility of the national governments whose prime function has been of ensuring security of supply. During the fuel crisis of the 1970s, many countries began exploring alternative sources of energy. The international community's first major attempt to develop a strategy for the use of alternative fuels was the 1981 UN General

Assembly Resolution A/RES/36/193 on the outcomes of the UN Conference on New and Renewable Sources of Energy. In this Resolution, the UN adopted the "Nairobi Programme of Action for the Development and Utilization of New and Renewable Sources of Energy," which addresses the need for an intergovernmental body, secretariat support, coordination within the UN system, regional and sub-regional action, cooperation among developing countries, and the mobilization of financial resources for new and renewable sources of energy. However, it was only in 1992 following the UN Conference on Environment and Development (UNCED) that renewable energy issues began to feature more prominently on the international environment and development agenda.

There is, as such, no global energy policy as, of now. However, the process of internationalization of energy policy has started with the global climate change concerns, increased trade in energy goods and services, and issues related to expansion of energy supply for economic and political reasons. At present there is a phase where the picture is not clear as to how and when, if at all possible, will a global energy policy take shape. However, the trends and common overlapping interests indicate that the world is moving towards a regime which will regulate or govern national energy choices in one or another way. A coherent global energy policy will look for a coordinated approach and linkage with other policies that govern transport, environment and development.

The following paragraphs provide an overview of the current policy trends and the international organizations involved in energy policy and programmes.

1.3.1 Developed Countries

1.3.1.1 Energy Policy Objectives

The underlying objective of energy policies of any industrialized (developed) country is to provide and maintain the security of supply. Alongside, equally important considerations are cost-effectiveness and low environmental impact. Although energy is not a State subject in almost all industrialized countries, there is considerable State intervention on the markets in the form of regulations to control prices and protect or otherwise monopolies. In addition, a decision to extract domestic energy resources, mainly fossil fuels, is taken after due consideration of the issues relating to long-term energy security and to ensure a large degree of energy independence, particularly during crises periods. All these factors contribute towards a heterogeneous mix of primary energy sources in these countries. This apart, steep hike in oil prices in 1970s led to strategic import and thrives towards becoming independent of energy imports, including developing new energy sources.

In order to provide energy at the lowest possible price, among other things, the general practice adopted is to establish security reserves. Such energy reserves particularly in respect of coal and gas, serve the twin purpose of ensuring supply and stabilizing market prices globally. In respect of grid based energy supply, liberalization of their electricity and gas markets is considered a viable option to attain greater economic efficiency and lower prices through competition.

1.3.1.2 Energy Related Environmental Concerns

One of the most important factor that influences the energy policy decisions in industrialized countries is environmental considerations. Whereas up to 1980s or beginning of 1990s reduction of localized pollution and NOx emissions due to energy extraction, processing and utilization was the major environmental issue that led to adoption of these government standards and policies for clean air at the local levels. However, with adoption of the United Nations Framework Convention on Climate Change (UNFCCC) in 1994 and subsequently the Kyoto Protocol in 1997, climate change abatement strategy became an important factor in the energy policies of these countries. As energy utilization is mainly responsible for emissions, improving efficiencies of the existing energy systems and exploring new and renewable sources of energy has been considered an important part of the energy policies in all developed countries, including US which has not ratified Kyoto Protocol and is out of the commitments.

1.3.1.3 Energy Charter Treaty

The Energy Charter Treaty (ECT) is an international treaty dealing explicitly with cross-border cooperation in the energy sector. The aim of the ECT is to promote economic growth through the liberalization of investment and trade. To this end, it extends the General Agreement on Trade and Tariff (GATT) rules to the energy sector. Minimum standards were agreed for foreign investment and energy transport. The Treaty, which evolved from the 1991 European Energy Charter, came into force in 1998. More than 60 countries, mainly from Europe and Japan and Central Asia have signed or acceded to the Treaty.

1.3.2 Developing Countries

The developing countries, although having individual policies, are not pursuing any unified energy policy and as such there are no defined trends in energy policy of these countries. As a group they are struggling with energy scarcity regimes. Around 1.64 billion people in these countries, which constitute around 30 per cent of world population, are still not having access to modern energy services.[10] Although there are substantial disparities between continents, countries, regions, urban and rural areas, energy structures and services are fast developing alongside the high economic growth to cater to the development aspirations of the people. Increasing urbanization and population growth are leading to an increase in the share of commercial energy. Rising from very low per capita income, in most of the countries, including India and China, investments in the energy sector are primarily government driven with heavy subsidies to the consumers. Although efforts are being made by fast growing economies towards liberalization and privatization of energy markets, the success horizon is still distant. In view of increasing energy demand the larger countries like India and China have started working towards long term energy security issues including development and use of renewable energy technologies.

1.3.3 Renewable Energy Policies

In view of energy security, access to energy and climate change concern, both developed and developing countries have started looking into renewable energy options. The programmes and policies of major developed and developing countries have been summarized in Chapter 3 and Chapter 4.

1.3.4 UN Initiatives

At the operational level, various UN agencies, including UNEP, UNESCO and UNDP, are involved in practical energy projects, *e.g.* to promote renewables and increase energy efficiency. UNEP's activities aim to increase the use of renewables and improve the efficiency of existing energy systems. At the political level, UNEP aims to incorporate environmental aspects into energy policy and improve analysis and planning in the energy sector. With its Energy and Atmosphere Programme, UNDP is pursuing an integrated development strategy which is designed to take into account many social, economic and ecological aspects of energy policy. The programme has been tasked with promoting sustainable energy policies and carrying forward the implementation of UNDP's energy programmes. The other international agencies which play a role in sustainable energy policy include the UN Regional Economic Commissions, the UN Committee on Energy and Natural Resources for Development, the United Nations Industrial Development Organization (UNIDO), the World Health Organization (WHO), the World Meteorological Organization (WMO), the United Nations Population Fund (UNFPA), and the Food and Agriculture Organization (FAO).

1.3.4.1 UN Conference on Environment and Development

At the United Nations Conference on Environment and Development (UNCED) in Rio de Janeiro in 1992, targets in the twin areas of human development and environmental protection were set out in Agenda 21 and the Rio Declaration. In Agenda 21 energy use was mentioned with reference to protection of the atmosphere, promoting sustainable agriculture and rural development, and promoting sustainable human settlement development. Chapter 9 of Agenda 21, on protecting the atmosphere, notes that much of the world's energy is currently produced and consumed in an unsustainable manner. It recognizes that the need to control atmospheric emissions of greenhouse gases and other substances will increasingly need to be based on efficiency in energy production, transmission, distribution and consumption, and a growing reliance on environmentally sound energy systems, particularly new and renewable sources of energy. The chapter also calls on governments and other stakeholders to, *inter alia* promote research, development, transfer and use of technologies and practices for environmentally sound energy systems, with particular attention to developing countries; review current energy supply mixes to determine how the contribution of environmentally sound energy systems could be increased in an economically efficient manner; examine and implement measures to overcome barriers; and coordinate energy plans regionally and sub-regionally.

1.3.4.2 UN Educational, Scientific and Cultural Organization

As a follow-up to UNCED, the UN Educational, Scientific and Cultural Organization (UNESCO) organized a High-level Expert Meeting in Paris in 1993, which launched the preparatory process for a world summit on the wider use of renewable energy. The World Solar Summit Process (WSSP) extended over a period of three years and included a series of expert and ministerial-level regional meetings. In 1994, the Executive Board of UNESCO approved the creation of a World Solar Commission with the mandate to provide advice on measures to reinforce global and regional cooperation for the promotion of renewable energy. The WSSP, which convened in Harare, Zimbabwe in September 1996, adopted the "Harare Declaration on Solar Energy and Sustainable

Development" and launched the World Solar Programme 1996-2005, a ten-year programme for the promotion of renewable energy. However, in the absence of a financial mechanism linked to it, WSSP did not make any headway. At present, UNESCO's activities mainly focus on information dissemination and training on different aspects of utilization of renewable energy in developing countries through the Global Renewable Energy Education and Training (GREET) and International Renewable Energy Information and Communication System (IREICS) programmes.

1.3.4.3 UN Commission on Sustainable Development

UN Commission on Sustainable Development (CSD) is another UN mechanism that helps in developing a consensus towards clean energy and set the agenda for discussion. The ninth session of the UN Commission on Sustainable Development (CSD) adopted a number of global energy policy recommendations in the publication 'Energy for Sustainable Development: A Policy Agenda'. They covered a wide range of issues including access to energy services, energy efficiency, renewables, next-generation fossil fuel technologies, nuclear power, rural energy systems, and transport. It identified the major challenge for both developed and developing countries as the development, utilization and dissemination of renewable energy technologies on a scale wide enough to contribute significantly to energy for sustainable development. The Commission, among others, recommended for developing and implementing appropriate national, regional and international policies and measures to create an enabling environment for the development, utilization and distribution of renewable energy sources.

1.3.4.4 World Summit on Sustainable Development

At the World Summit on Sustainable Development (WSSD) in 2002, energy was included as a separate agenda item, with the focus on renewables, access to energy services, organization of energy markets, and energy efficiency. One of the major outcomes of the WSSD was the adoption of the Johannesburg Plan of Implementation (JPOI), which addresses renewable energy in several of its Chapters. Regarding poverty eradication (JPOI Chapter II), governments agreed to improve access to reliable and affordable energy services for sustainable development, so as to facilitate the achievement of the Millennium Development Goals (MDGs). These included actions to:

- Improve access to reliable, affordable, economically viable, socially acceptable and environmentally sound energy services and resources through various means, such as enhanced rural electrification and decentralized energy systems, and increased use of renewables;
- Improve access to modern biomass technologies and fuelwood sources and supplies, and commercialize biomass operations; and
- Promote the sustainable use of biomass and other forms of renewable energy through improved patterns of use.

Regarding sustainable consumption and production patterns (JPOI Chapter III), governments agreed to increase substantially the global share of renewable energy sources, with the objective of increasing the contribution of renewable energy to total energy supply with a sense of urgency. They recognized the role of national and voluntary regional targets and initiatives, and the need to ensure that energy policies support developing countries' efforts to eradicate poverty. Apart from other things, they also agreed to develop and disseminate alternative energy technologies with the aim of giving a greater share of the energy mix to renewable energy; combine the increased use of renewable energy resources, more efficient use of energy, and greater reliance on advanced energy technologies; and develop and utilize indigenous energy sources and infrastructures for local use and promote rural community participation in the development and utilization of renewable energy technologies.

1.3.4.5 UN Framework Convention on Climate Change

The United Nations Framework Convention on Climate Change (UNFCCC) defines objectives and principles, bodies and mechanisms for international climate protection policy. It entered into force in 1994 and has been ratified by 184 states around the world, including all the major industrialized and developing countries. Climate policy entails, above all, a radical reduction in GHG emissions worldwide. This can only be achieved through the comprehensive restructuring of global energy systems. Here, UNFCCC has a key role to play in international energy policy. It is the driving force and the most significant International forum where States can discuss the interface between environmental and energy policy and adopts major decisions. The Kyoto Protocol to the UNFCCC was adopted in 1997 by more than 160 nations. The Protocol sets out specific reduction commitments for a defined group of greenhouse gases. It commits the industrialized countries to reducing greenhouse gas emissions by at least 5 per cent by the 2008–2012 period against the baseline year of 1990. Annex I of the Protocol stipulates a precise reduction target for each industrialized country (EU –8 per cent, USA –7 per cent, Japan –6 per cent, Australia +8 per cent, and Russia 0 per cent). The protocol entered into force on February 16, 2005, following ratification by Russia on November 18, 2004. As of December 2007, a total of 170 countries and other governmental entities have ratified the agreement (representing over 61.6 per cent of emissions from UNFCCC countries). Although United States are of the major GHG emitters has signed the Kyoto Protocol but has yet not ratified it. Other countries, like India and China, which have ratified the protocol, are not required to reduce carbon emissions under the present agreement despite their relatively large populations.

The protocol allows the industrialized countries some flexibility in implementing their commitments through joint implementation projects and emission trading between the countries having emissions reduction targets and Clean Development Mechanism (CDM) to earn carbon credits from carbon neutral or less carbon emitting projects with respect to the on-going scenario in developing countries while complying with specific rules and modalities as set by CDM Executive Board. The CDM project portfolio and market are growing rapidly. Apparently, international climate policy has had a positive impact on national, regional and global energy policy. An analysis of CDM project portfolio has been made in Chapter 5.

1.3.5 G-8

In July 2000, the leaders of the eight major industrialized countries (G-8) and the President of the European Commission (EC) met in Okinawa, Japan for the G-8's 26th Summit. At the Summit, the G-8 established the G-8 Renewable Energy Task Force to identify actions to promote a change in the supply, distribution and use of renewable energy in developing countries. The Task Force in 2001, concluded that renewable energy resources can sharply reduce local, regional and global environmental impacts as well as energy security risks. It can also, in some circumstances, reduce energy costs for consumers. The Task Force suggested that concerted action by the G-8, other countries, the private sector, and international financial institutions to implement the Task Force's recommendations over the next decade could result in various positive outcomes, including electricity access from renewable sources for up to 300 million people in rural areas of developing countries, and service for up to 500 million people connected to electricity grids worldwide.

1.3.6 Renewable Energy Policy Network for 21st Century

The Government of the Federal Republic of Germany hosted the International Conference on Renewable Energies during 1-4 June 2004 at Bonn, Germany. The major outcome of the conference was the International Action Programme, through which many of the national governments, international organizations etc. committed themselves towards accelerated use of renewable energies with defined goals. The formation of Renewable Energy Policy Network for 21st Century (REN 21) as a global policy network is an outcome of this conference. REN 21's stated goal is to bolster policy development for the rapid expansion of renewable energies in developing and industrialised economies.

1.3.7 Energy and WTO

With the conclusion of the Uruguay Round in 1994, a number of energy carriers, especially coal, gas and oil and partly also electricity, were integrated into the world trade regime. The World Trade Organization (WTO) secretariat views subsidies on energy provision and consumption as the most important barrier to liberalization of the trade in energy. The full integration of this sector into the WTO rules and the application of the WTO Agreement on Subsidies and Countervailing Measures may result in reduced subsidies and thus make a major contribution to climate protection at the same time. According to the WTO, the dismantling of all subsidies by 2010—supported by appropriate flanking measures in the environmental policy field—could prevent around 6 per cent of CO_2 emissions worldwide.[11] However, full integration of the energy sector into the WTO may restrict the scope for subsidies and other measures for renewable energy technologies.

1.3.8 International Financing Structures for Energy

1.3.8.1 Global Environment Facility

The Global Environment Facility (GEF), established in 1991, helps developing countries fund projects and programmes that protect the global environment. GEF grants support projects related to biodiversity, climate change, international water, land degradation, the ozone layer, and persistent organic pollutants. GEF has three Implementing Agencies—World Bank, UNDP and UNEP. GEF projects in climate change help developing countries and economies in transition to contribute to the overall objective of the United Nations Framework Convention on Climate Change (UNFCCC). The projects support measures that minimize climate change damage by reducing the risk, or the adverse effects, of climate change. As the financial mechanism of the Climate Convention, GEF allocates and disburses about US $250 million per year for projects in energy efficiency, renewable energies, and sustainable transportation. The GEF also manages two special funds under the UNFCCC—the Least Developed Countries Fund and the Special Climate Change Fund—as well as the Adaptation Fund under the Kyoto Protocol. The desired outcome from GEF interventions is to create favourable conditions for market development in terms of policy, finance, business models, information and technology.

1.3.8.2 World Bank

Alongside GEF, the World Bank is the most important provider of energy policy financing, especially for developing countries. Since the World Bank's establishment, the energy sector has been a key area for lending, with up to 25 per cent of the World Bank's total lending occurring in this sector for many years. In April 2006 the World Bank put on the public domain a paper titled "Clean Energy and Development: Towards an Investment Framework," which elaborate upon the Bank's strategy towards cleaner energy sources. It observes that developing countries and industrializing economies face major energy sector challenges namely: (a) improving energy services for economic growth; (b) providing energy services for a better quality of life; (c) reducing local and regional pollution from energy production and use, and (d) strengthening energy security for macroeconomic stability. The Bank has detailed its approach to energy sector interventions which envisions efficient supply and use of energy that strengthens the economic growth of developing and transition economies by:

- (i) freeing consumers from frequent and prolonged power cuts and liquid fuel shortages;
- (ii) giving industrial enterprises a choice of suppliers providing reliable energy services;
- (iii) allowing private ownership and financing a dominant role in energy supply;

(iv) ensuring that regulators operating in an objective, transparent, and non-discriminatory manner oversee natural monopolies and promote competition in the energy sector;
(v) reducing the average intensity of carbon dioxide emissions from energy production; and
(vi) reducing the average energy consumption per unit of GDP.

In order to address these issues the Bank has proposed additional financing options, on the assumption that even with an improved regulatory environment and the use of policy and political risk mitigation instruments, the challenge of financing incremental costs and reducing technology risks will be significant. These options are: (a) Clean Energy Financing Vehicle (CEFV); (b) Power Rehabilitation Financing Facility; (c) Project development fund; and (d) Venture capital funds for technology adoption.

1.3.9 Towards a Global Framework

While a common basis for an international energy framework is being attempted through various measures, the shape of the things to come is yet not clear. However, with a fair degree of certainty it could be concluded that over the years energy utilization will be governed by one or another kind of direct or indirect policy regime. The main drivers for such a regime will be climate change and energy security concern. In such a development renewable energy, without doubt, will remain central to any discussion and get focused attention.

NOTES

1. Nebojša Nakiæenoviæ, International Institute for Applied Systems Analysis (IIASA) and Vienna University of Technology, A presentation on 'Emission Scenario Premier" made at Aspen Global Change Institute, Aspen, United States on 7 July 2003: http://www.agci.org/energyPPT/naki-aspen-2jul03.ppt assessed on 6 March 2007.

2. Nakicenovic, N Grubler, A and McDonald, A (1998) Global Energy Perspective. Cambridge University Press, Cambridge and New York.
3. Generally referred to developed countries which constitute around 20% of world population. Its member countries are Australia, Austria, Belgium, Canada, the Czech Republic, Denmark, Finland, France, Germany, Greece, Hungary, Ireland, Italy, Japan, the Republic of Korea, Luxembourg, the Netherlands, New Zealand, Norway, Portugal, Spain, Sweden, Switzerland, Turkey, the United Kingdom, and the United States.
4. Key Energy Statistics, 2007: International Energy Agency.
5. Global Energy Perspective (1998); Cambridge University Press, Cambridge and New York (Nakicenovic N., Grubler A. and McDonald A.).
6. http://www.grida.no/climate/ipcc/emission/072.htm#tbl36
7. The primary tool used for the analysis of the ACT scenarios is the IEA Energy Technology Perspectives model (ETP). This global 15-region model permits the analysis of fuel and technology choices throughout the energy system, from energy extraction through fuel conversion and electricity generation to end-use. The model's detailed representation of technology options includes about 1000 individual technologies. The ETP model belongs to the MARKAL family of bottom-up modelling tools.
8. Individual scenarios which differ in terms of assumptions for key technology areas are called:
   ACT Map: Relatively optimistic across all technology areas and 2.0 per cent per year global improvement for end-use efficiency,
   ACT No CCS: CO_2 capture and storage (CCS) technologies will not be available,
   ACT Low Renewables: Slower cost reductions for renewable energy technologies,
   ACT Low Nuclear energy: Lower public acceptance for nuclear energy,
   ACT Low efficiency: 1.7 per cent per year global improvement for end-use efficiency,
   TECH Plus: Stronger cost reductions for nuclear and renewable energy; Breakthrough for Hydrogen and fuel cell technologies; Stronger cost reductions and improved feedstock availability for advanced biofuels.

9. The key features of the Map scenario include: (a) barriers to the capture and storage of CO_2 are overcome, although costs remain high; (b) cost reductions for renewable energy technologies, such as wind and solar, continue with increasing deployment; (c) expansion of nuclear power; (d) progress in energy efficiency; (e) biofuels become an increasingly viable alternative; (f) significant progress in costs of hydrogen fuel-cell vehicles cost reduction.
10. International Energy Agency, 2002.
11. WTO 2001; International Trade Statistics 2001, WTO Geneva.

Chapter 2

Renewable Energy Technologies: Present Status & Future Projections

As per UN Commission on Sustainable Development the primary energy derived from geothermal, hydro, solar, tide, wind and wave power, and biofuels, such as fuelwood, bagasse, charcoal, animal and vegetal wastes, and other (industrial and municipal) wastes are renewable energy. Further, the European Union under its directive 2001/77/EC dated 27 September, 2001 has defined renewable energy sources as non-fossil energy sources such as wind, solar, geothermal, wave, tidal, hydro-power, biomass, land fill gas, sewage treatment plant gas and biogas. The Political Declaration of the International Conference for Renewable Energies held in Bonn in June, 2004 defines renewable energy sources as solar energy, wind energy, hydropower, biomass energy including biofuels, and geothermal energy. Renewable energy encompasses a wide variety of resources and technologies. Theoretically, the renewable resource are very large. Many of the technologies are technologically viable and many are well proven, but with exceptions their costs are high relative to fossil fuels. As such, renewable energy technologies are still evolving in terms of technological maturity, cost competitiveness and face one or another kind of barriers. Some technologies are not yet ready for the market.

Research and Development (R&D) and technology learning are necessary for further cost reduction of renewable energy technologies. Technology learning is a process in which production costs decrease and technical performance increases as cumulative installed capacity rises and technologies are made available at lower costs.

Table 2.1 Cost of Renewable Power over a Period of Time

Technology	*Learning Rate (per cent)*	*Investment Cost (US$/KW)*			*Production Cost (US$/MWh)*		
		2005	2030	2050	2005	2030	2050
Biomass	5	1000-2500	950-1900	900-1800	31-103	30-96	29-94
Solar PV	18	3750-3850	1400-1500	1000-1100	178-542	70-325	<60-290
Solar Thermal	5	2000-2300	1700-1900	1600-1800	105-230	87-190	<60-175
Wind on shore	5	900-1100	800-900	750-900	42-221	36-208	35-205
Wind Off shore	5	1500-2500	1500-1900	1400-1800	66-217	62-184	60-180

The prospect that a given technology will be produced and sold on the market can stimulate private industry R&D and improvements in the manufacturing process. Feedback from the market may suggest avenues for improving a technology, reducing its costs or tailoring some of its features to consumers' needs. The rate of technology improvement is therefore usually a function of the rate of technology adoption. The projected cost and investments for various renewable power technologies over a period of time using the past learning rates (percentage cost reduction for each doubling of installed capacity) of the respctive technologies using 10 per cent discount rate are summed up in Table 2.1.[1]

As such technology development has various stages, R&D, demonstration, deployment and competitive. The R&D stage is to overcome technical barriers and to reduce costs. At this stage the outcome still remains uncertain and requires technology demonstration. As the technology is yet to be proven, public funds are generally needed to finance demonstration projects. Once technology is successfully demonstrated, it requires creation of volumes by large scale application through policy and other kinds of supports. Increasing application result in gradually decreasing costs. The logical end of this process is the technology becoming cost competitive in some or all markets and may still require sustained support to make it a baseline technological option. Once a technology has been proved as viable option for dissemination, other barriers such as public acceptance, availability of finance, information, incentives, etc. are also required to achieve full market application.

2.1 Renewable Energy Technologies Assessment

A review of major renewable energy technologies—solar photovoltaic, solar thermal power, wind electric, biomass power, biofuel and hydrogen have been attempted in the following sections. Although there exist many other renewable technologies such as ocean energy, geothermal and hydro power these have not been covered in view of their level of maturity and/or their perceived role in next 50 years. The underlying objective is provide an overview of the present level and future projections in terms of cost reduction and technological improvements.

2.1.1 Solar Photovoltaics

2.1.1.1 Technology

Photovoltaic (PV) cells are based on semiconductors and convert light directly into electricity. They are electrically connected in series and encapsulated within modules. Photovoltaic modules can be connected to module arrays to provide power from few watts to several hundred watts that can be combined into larger power arrays. These systems are connected to consumers or to the grid via electronics. Solar-photovoltaic technologies include off-grid and on-grid applications. Rechargeable battery is usually integrated in systems that are not connected to the grid. Since the development of the first solar cell in 1954, the dominant raw material for solar cell production has been crystalline silicon. The cost of PV has declined by a factor of 100 since the 1950s, more than any other energy technology in the period. Markets for PV are expanding rapidly, recently in excess of 30 per cent per year.[2]

The overall efficiency of systems available on the market depends on the type of cell. Crystalline silicon has been the most important PV technology so far. The first thin film PV device, an amorphous silicon (a-Si) module, was developed in the 1980s. More recently, other semiconductors, including cadmium telluride (CdTe) and copper indium diselenide (CISD), have been used in industrial module production. The potential for thin-film modules is considered very high, but so far their diffusion has been limited by their high cost. Currently achievable efficiencies of typical technologies are summarized in Table 2.2.[3]

Table 2.2 Solar Cell Efficiencies

Cell technology	*Wafer technology/Thin film technology*	*Efficiency in the laboratory [%]*	*Efficiency of the module [%] (The figures in brackets relate to values from pilot production or initial commercial production)*
Monocrystalline silicon	Wafer technology	24.7	13–15
Multicrystalline silicon	Wafer technology	19.8	12–14
Concentrator cells (incl. tandem and triple)	Wafer technology and Thin film technology	33.5	(25)
Crystalline thin film silicon cell	Thin film technology	19.2	
Organic and dye solar cells	Thin film technology	2–11	

The efficiency of photovoltaic devices has been increasing steadily. However, current technologies all lie on a relatively common cost/watt scale. Large-grain, pure materials, with a long lifetime capable of making efficient solar cells, are costly to produce. Alternatively, cheaper materials with smaller grain sizes have grain boundaries that act as recombination sites, resulting in inefficient solar cells. A similar trade-off is found for organic (plastic) photovoltaics. If pure inorganic single-crystal materials, like silicon and GaAs, are replaced with much cheaper organic materials, the materials are inherently disordered and therefore are cheaper but more inefficient. The net result is that one can expect anywhere on this cost/watt trade-off scale, but nevertheless end within a factor of 20 per cent.

2.1.1.2 Cost Projection

Conversion efficiencies of PV modules have been improving at a constant pace through the use of new materials and cell designs. The cost of PV electricity depends on the level of insolation. About half of potential future cost decreases for PV is expected to result from R&D directed towards improving materials, processes, conversion efficiency and design. Substantial cost reductions can also be gained through increased manufacturing volume and economies of scale.

Over the next decade, thinfilm technologies are expected to display their potential for cost reduction and improved performance. The cost for PV modules is about the same as for all other needed system components together. A number of scenarios have been attempted by various research groups for PV technologies development and future cost reduction. A number of technologies are in the commercial stage; many others are still in the pil . manufacturing or even laboratory phase. The most possible scenario is of different technologies continuing to co-exist for different applications. Most scenarios indicate towards diversification of PV technologies, improved efficiencies and cost reduction. These are summed up in Table 2.3.[4]

Table 2.3 Future Development of Photovoltaics

	2000	*2020*	*2050*
Main, market-dominating technologies	Crystalline silicon solar cells	Crystalline Si and thin film solar cells. Tandem solar cells for PV power plants with optical concentration. Organic and dye solar cells	Thin film solar cells (incl. Si), tandem solar cells, organic and dye solar cells, new concepts
Module efficiency [%]	14–15	Si wafer module: 18–20 Thin film: 15 Organic, etc.: 10 Tandem: 40	no info available
Costs [US$/kWh]	~0.8 (Location with 1000 full load hours)	~0.18 (Location with 1300 full load hours)	~0.08 (Location with 1300 full load hours)

2.1.1.3 R & D Focus

There appears to be twin approaches governed by specific characteristics of the respective technologies. Crystalline silicon technology is comparatively well established and the long-term strategy would be to bring down costs by evolutionary improvements based on further research and technology development and also economies of scale. The thin film and new concepts such as dye-sensitised cells, organic cells or nanotechnology-related concepts are yet to have market penetration and so far been limited to laboratory or demonstration stages. The future research efforts will be aimed to achieve a major breakthrough in cost reduction with the introduction of such new concepts and materials. As such cost reduction is a key issue for PV, as costs are still relatively high compared to those for other electricity generation technologies. The key drivers for future reductions in PV electricity costs include decreasing cell costs while maintaining or increasing efficiency; increasing module lifespan; reducing specific module costs through higher cell efficiencies and more advanced encapsulation techniques; and reducing specific balance-of-system costs through higher module efficiencies and improved electronics. The cost reduction would be possible through lower specific material consumption, more efficient production schemes and new cell concepts. In addition, development of standards for all the main system components, as well as certification or qualifications for designers and installers would be important for large scale application.

2.1.2 *Concentrating Solar Power*

2.1.2.1 Technology

In solar thermal power plants, direct sunlight is concentrated onto an absorber *via* optical elements. The absorbed radiation energy heats a heat transfer medium. This heat energy can subsequently be used for driving largely conventional engines such as steam turbines or Stirling motors. Solar thermal power plants are therefore closely associated with classic power plant technology, with solar energy replacing fossil fuels. The systems use either thermal storage or back-up fuels to offset solar intermittency and thus to increase the commercial value of the energy produced. There are three types of concentrating solar power (CSP) technology, parabolic trough power plants, solar power towers and parabolic dish power plants.

As concentrating solar power uses direct sunlight, the best conditions for this technology are in arid or semi-arid climates, including Southern Europe, North and Southern Africa, the Middle East, Western India, Western Australia, the Andean Plateau, North-eastern Brazil, Northern Mexico and the Southwest United States. Solar Thermal Power Plants can also be operated in combined heat and power mode with resultant very high efficiencies. The projections indicate that in the long term, the concepts of parabolic troughs or solar power towers appear more promising than smaller parabolic dish systems. In sunny regions, parabolic troughs and solar power towers are currently by far the most cost-effective option for generating electricity using solar energy in large power plants. Due to their different features, it is expected that both technology options will be used widely, depending on the underlying conditions.

Solar thermal power plants were initially setup in California state of USA in 1980s and early 90s. After a gap of around one decade solar plant projects were being developed in Spain and in the United States, under the respective provisions of Renewable Portfolio Standard.

2.1.2.2 Cost Projection

The power generation cost from CSP technologies is expected to fall by half by 2020 and to one-third of the present cost by 2050 (Table 2.4).[5] CSP technology is still expensive to compete with conventional energy technologies without subsidies. The ongoing research efforts are aiming at reducing the cost of CSP systems.

Table 2.4 Solar Thermal Power Technologies

	2000	*2020*	*2050*
Main technologies	Parabolic troughs	Parabolic troughs and solar power tower	Parabolic troughs and solar power tower
Electrical efficiency [%]	14	20–25	25–30
Costs [US$/kWh]	0.18–0.26	~0.09–0.18	~0.08

Most of the technological components of solar thermal-power plant systems still need improvement. Improved manufacturing technologies are needed to reduce the cost of key components. Attaining higher operating temperatures and efficiencies in parabolic trough systems by using steam as a heat transfer medium and improved selective absorbers and also advanced storage systems are major research areas. Another research and development area is producing reflector materials with good outdoor durability, high solar reflectivity, good mechanical resistance to withstand periodical washing and low soiling co-efficient. In addition to improved technology, the widespread application of CSP plants is hindered by the heavy investment required for large centralised power plants. As such, building a sustainable market for CSP requires taking the lessons learned in the countries where CSP application is successful and transferring them to others. This would promote faster market growth, attract larger global companies and lead to costs that are increasingly competitive with conventional sources.

2.1.3 Low-Temperature Solar Thermal Systems

2.1.3.1 Technology

Low-temperature solar thermal systems have typical applications in domestic or industrial hot water, space heating, or process heat etc. with the required temperature increasing in the above order. The higher the required temperature level, the more complex the collector, since more effort is required to avoid heat losses. There are three basic types of thermal collector, (a) collectors with unglazed absorber: *for moderate temperatures,* *as w*ater flows through black plastic mats without cover; (b) *flat plate collectors: for* significantly higher temperatures, as the absorber is covered with a glass pane, and at the back it is insulated; and (c) *evacuated tube collectors: for* good efficiencies and high temperatures as with this type of collector, vacuum insulation almost completely eliminates heat losses through conduction and convection from the absorbers, which are located within glass tubes. Each type of collectors have specific advantages for particular niche application.

Solar heating and cooling is already commercial in many applications, particularly for industrial process heat requirements. However, its application is far below its potential.

2.1.3.2 Cost Projections

The cost of solar water heating systems depends on size and location. Solar hot water heating systems generally cost between US$ 1-2 per watt of capacity, with the cost of energy supplied varying depending on the location and sunshine hours per year. Although the systems have already been delivering energy at competitive prices it is estimated that further cost reductions in the range of 40 to 50 per cent are achievable through further development.[6]

2.1.3.3 R&D Focus

Standardization of systems, sub-systems and components and also system materials will be the major areas for future research. In solar heat application area the major challenges lies in creating conditions for large scale application. Another major area for research is application of solar energy for cooling applications. There are two ways, the first using solar electricity to operate a conventional compression-type refrigerating machine and the second is to use solar heat to drive thermodynamic cooling processes via sorption or adsorption techniques. Development of hybrid cooling systems utilizing solar and conventional energy would be an ideal option to minimize consumption of primary energy required for the operation of a conventional system.

2.1.4 Bioenergy

2.1.4.1 Technology

Bioenergy technology converts the chemical energy stored in organic matter into heat and power. It encompasses a broad range of solid, gaseous, and liquid fuels that result from living organisms or from the wastes and by-products of human activities. The technology choice for converting biomass into useful energy for producing energy for heat or power is determined by the biomass feedstock and the cost to be incurred on the energy generation is set by feedstock costs at source and also cost of transportation to the end use point. There are three major methods, combustion, gasification and anaerobic digestion for converting biomass into useful energy.

Biomass combustion technologies convert biomass fuels into several forms of useful energy including hot air, steam or power generation. A biomass fired boiler generates high pressure steam by direct combustion of biomass in a boiler. The steam then passes through a steam turbine to produce electricity. Bio-electric plants are an order of magnitude smaller than coal-fired plants based on similar technology. Average plant efficiencies are in the range of 30 per cent in MW range capacities. However, higher efficiencies could also be obtained with use of wood chips at high steam temperature. Partial substitution of coal by biomass is possible and efficiencies on the order of 35 to 45 per cent can be achieved in modern plants.

Biomass gasification is the process through which solid biomass material is subjected to partial combustion in the presence of a limited supply of air. The ultimate product is a combustible gas mixture known as producer gas. The combustion of biomass takes place in a closed vessel, normally cylindrical in shape, called a gasifier. Producer gas typically contains nitrogen (50-54%) carbon dioxide (9-11%), methane (2-3%), carbon mono oxide (20-22%) and hydrogen (12-15%). Producer gas has relatively low thermal value, ranging from 1000-1100 kcal/m^3 depending upon the type of biomass used. The gas can be used to drive engines, steam or gas turbines.

Table 2.5 Global Potential Biomass Energy Supply

Biomass category	*Main assumptions and remarks*	*Potential bio energy supply in 2050 in Mtoe*
Bioenergy farming on current agricultural land	Potential land surplus: 0-4 Gha (average: 1-2 Gha). A large surplus requires adoption of intensive agricultural production systems. Higher yields are more likely where soil quality is good. Productivity of 8-12 dry tonnes/ha/year assumed. (Heat content of 19 GJ/tonne dry matter is assumed)	0—16700 (Average: 2 400—7200)
Bioenergy production on marginal land	Maximum land surface of 1.7 Gha could be used. Low productivity of 2-5 dry tonnes/ha/year assumed (Heat content of 19 GJ/tonne dry matter is assumed) Poor economics or competition with food production could limit availability.	1 400—3600
Residues from agriculture	Potential depends on yields and total agricultural land area in use as well as type of production system (extensive production systems require re-use of residues to maintain soil fertility).	400—1700
Forest residues	The sustainable energy potential of the world's forests is uncertain. Low value corresponds to sustainable forest management and high value to technical potential. Estimates include processing residues.	700—3600
Dung	Low estimate based on global current use, high estimate on technical potential. Collection rates are uncertain.	100—1300
Organic wastes	Estimates dependent on assumed rates of economic development and consumption of biomass generally. Higher values correspond to more intensive use of biomass.	100—1000
Total	Low estimate based on no land available for energy farming and use of residues only. High estimate based on intensive agriculture concentrated on high-quality soils.	1000—26200 (Average: 6000—11900)

Through biological process, organic waste can be partly converted into a gas containing primarily methane as an energy carrier. Biogas could be used for various purposes including cooking, heating and electricity generation. After purification, the gas can be used for production of transport fuels. While liquid state technologies are currently the most common, recently developed solid-state fermentation technologies are also widely used. Anaerobic digestion technologies are very reliable, but they are site-specific and their capacity for scaling-up is limited.

There exist various estimates about bio-energy potential in long term. IEA has estimated the potential bioenergy supply in 2050 between 6000-11000 Mtoe (Table 2.5).[7]

2.1.4.2 Cost Projection

The cost of producing electricity from solid biomass depends on the technology, fuel cost and fuel quality. The average estimated additional cost of biomass co-firing with coal is in the range of US $ 50—250 per kW of biomass capacity. Very low generation costs can be achieved with co-firing in situations where little additional investment is needed and biomass residues are available for free. Estimates indicate that generation costs could be in the range from US$ 0.10 to US$ 0.15 per kWh in innovative gasification plants. Biomass integrated gasifier/gas turbine plants have long-term potential in terms of both efficiency and cost reduction. Larger plants require that biomass is transported greater distances to the generating station, and long transport routes make biomass less attractive in both economic and environmental terms. For even lower capacities, gasification will probably be combined with gas engines or turbines in combined heat and power units, replacing current steam processes.

2.1.4.3 R&D Focus

Adequate availability of biomass resources is considered as one of the major deterrent in accelerated penetration of biomass conversion technologies. In addition, the biomass resources have alternative competing uses that further restrict its utilization for energy generation. The sustainable use of biomass in the energy sector may have competition with food production and also in the developing countries context there are issues of use of biomass as fuels versus power generation.

The major challenge in biomass utilization technologies is conversion of solid biomass into liquid bio-fuel at economic costs. While biomass combustion plants are commercially available in various sizes, their efficiency could still be improved and their costs further reduced. Technology improvements are also needed in areas such as gasification, the development of plants that can use a variety of biomass feedstocks, polygeneration and co-firing using a range of feedstocks. Improvement in the efficiency of cogeneration units run on bagasse is another major area for action. There is also a need to work for standardisation of feedstocks and technologies that would ultimately lead to mass production and cost reduction. Further, gasification technologies need to demonstrate reliable commercial operation. The research areas that need attention include efficient tar removal and design standardization for multi fuel systems, multi-fuel co-firing with biomass and waste and flue gas cleaning requirements.

2.1.5 Biofuel

2.1.5.1 Description

The term biofuels in the present context means liquid fuels produced from biomass, and used in place of diesel, petrol or other fossil fuels, for transport, stationary, portable and other application. There are several types of biofuels and many different ways of producing them. Presently, almost all biofuels produced around the world are either ethanol or biodiesel. Ethanol is usually produced from sugar and starchy crops, such as cereals, while biodiesel is produced mainly from oil-seed crops. Each fuel has its own unique characteristics, advantages and drawbacks. Ethanol, in an almost water-free form, is usually blended with gasoline, either pure or in a derivative form, known as ethyl-tertiary-butyl-ether. Biodiesel can be used easily in most existing compression-ignition engines in its pure form or in virtually any blended ratio with conventional diesel fuel. Ethanol in a hydrous form (containing up to 5 per cent water) and biodiesel can be used unblended or in high-proportion blends only with modifications to the vehicle engine. Almost all biofuels are used in cars and trucks, though small quantities of ethanol are used for aviation purposes. Global production of biofuels was 20 Mtoe, or 643 thousand barrels per day (kb/d) in 2005 which was around 1 per cent of total road-transport fuel consumption in energy terms.[8]

Table 2.6 Biofuels Production 2005

	Ethanol		*Biodiesel*		*Total*	
	Mtoe	*kb/d*	*Mtoe*	*kb/d*	*Mtoe*	*kb/d*
United States	7.50	254	0.22	5	7.72	259
Canada	0.12	4	0.00	0	0.12	4
European Union	0.48	16	2.53	56	3.01	72
Brazil	8.17	277	0.05	1	8.22	278
China	0.51	17	negligible		0.51	17
India	0.15	5	negligible		0.15	5
World	17.07	579	2.91	64	19.98	643

IEA has projected that the world production of biofuels would grow at an average annual rate of growth of 6.3 per cent upto 2030. However, with favourable policy and support regime the growth could be 8.3 per cent. This translates into increase by 5-7 times from 2005 level.

2.1.5.2 Production Technologies

2.1.5.2.1 Ethanol

Conventional ethanol production technology involves fermenting sugar obtained directly from sugar cane or beet, or indirectly from the conversion of the starch contained in cereals. The ethanol produced is then distilled to produce a fuel-grade liquid. The ethanol production using starch crops starts with the separating, cleaning and milling of the starchy feedstock. The starch is converted into sugar in a high-temperature enzyme process. After this point, the process is similar to that of sugar crops, where the sugars are fermented into alcohol using yeasts and other microbes.

In the conventional grain-to-ethanol processes, only the starchy part of the plant is used for the production of fuel. These starchy parts, however, represent a fairly small percentage of total plant mass, leaving considerable fibrous remains (seed, husks and stalks). Much of current research is focused on innovative processes to also use these lignocellulosic materials to create fermentable sugars. Lignocellulosic feedstocks typically by weight contain 20 to 45 per cent of cellulose, 20 to 25 per cent of hemicellulose and other components including lignin. The ethanol from lignocellulosic feedstocks production process requires (a) cleaning and breaking down of the feedstock by a combination of physical and chemical processes at high temperatures; (b) separation of cellulose, hemicellulose and lignin through hydrolysis process and conversion of cellulose and hemicellulose to sugars; (c) conversion of the sugars into ethanol through fermentation; (d) separation of solid residue from the liquid solution containing ethanol; and (e) distillation and rectification of ethanol to fuel grade.

2.1.5.2.2 Biodiesel

Biodiesel is mainly produced from transesterification of vegetable oils. The process involves filtering the feedstock to remove water and contaminants and then mixing it with an alcohol (usually methanol) and a catalyst (usually sodium hydroxide or potassium hydroxide). This causes the oil molecules (triglycerides) to break apart and reform into biodiesel and glycerol, which are then separated from each other and purified. Biodiesel contains only about 90 per cent as much energy as conventional diesel, but its lubricity and higher Cetane number (a measure of the combustion quality of diesel under compression) mean that fuel economy is similar.

Conventional biodiesel crops include rapeseed, palm, sunflower, soya, jathropa and other oil-seed bearing crops. The major producers of soyabean are the United States and China. It has a multitude of uses including edible vegetable oil and proteins for livestock feed use. The main producers of sunflower seeds are Russia and Argentina and also widely cultivated in China, India, the United States and Europe. China, India, Europe and Canada are main producers of rapeseed oil and it is used for the production of edible oil in Asia, and elsewhere for the production of animal feed, vegetable oil and biodiesel. Palm plants are mainly grown in Malaysia and Indonesia.

In most of the developing countries conventional bio oil is used for human consumption and therefore the focus is to develop non-edible bio oil plants such as Jatropha, Pongamia etc. which can be grown on degraded lands. The purpose is to not to compete with land where food production would be possible.

Crops for biodiesel demand more than three times as much land as sugarcane used for ethanol to deliver the same amount of biofuel energy. Sunflower and rapeseed lead to much lower biofuel yields per hectare than those for ethanol. The typical yield of soybeans cultivated in Brazil and rapeseed cultivated in Europe and palm oil in Malayasia are 700, 1100 and 900 litres of diesel equivalent per hectare respectively.

Biomass gasification combined with Fischer-Tropsch (FT) synthesis is a process of converting solid biomass into liquid biofuel. In this process first the biomass is thermally converted into the synthetic gas and thereafter FT synthesis is used to convert the syngas into fuel. As yet, there is no commercial production of biofuels through asification, because of the high cost compared with conventional technologies.

Another approach is fast pyrolysis in which biomass is quickly heated to high temperatures in the absence of air and then cooled down, forming a liquid ("bio-oil") plus various solids and vapours. However, at present, fast pyrolysis is associated with unacceptable energy and financial costs.

2.1.5.3 Costs

The cost of producing biofuels depends on the type of feedstock and the conversion technology. Moreover, costs vary by region, depending on biomass yields, agricultural subsidies, land costs, labour costs and availability of capital. As such no single cost figure can be cited for producing biofuels.

2.1.5.3.1 Ethanol

Ethanol produced in Brazil from cane sugar costs about US$ 0.30 per litre of gasoline equivalent free-on-board (FOB). Ethanol produced in other world regions is likely to cost more around US$ 0.40—0.50 per litre of gasoline equivalent.

Maize-based ethanol produced in the United States is sold at about US$ 0.80 per litre of gasoline equivalent. Its production cost is estimated to be close to US$ 0.60 per litre. In Europe, production costs for ethanol derived from sugar beets are similar, while producing ethanol from wheat costs about US$ 0.10 to 0.15 more per litre. As the conversion technologies in this segment are comparatively mature, real term price reduction is unlikely.

Ethanol made from lignocellulosic biomass (*e.g.* grass, wood, and various agricultural and forestry wastes) takes advantage of significantly lower raw material cost, more plentiful and varied feedstocks, and expands the potential for ethanol to blend with and displace gasoline with a cleaner and renewable liquid fuel. Further, the benefits of developing biomass-to-ethanol technology is that it does not compete with food markets, and therefore no competition with food security. The present cost of lignocellulosic ethanol production has been estimated at around US$ 1.00 per litre of gasoline equivalent. Although there are no firm estimates about cost reductions, in the long term with better ethanol concentrations before the distillation and lower costs for enhanced enzymes and also improved separation techniques, the cost is expected to decrease to US$ 0.50 per litre. This apart, cost reduction is also expected from the upscaling of production facilities and lower feedstock cost for ethanol production units using biomass residues as feedstock.

2.1.5.3.2 Biodiesel

The current cost of conventional biodiesel production is estimated at just over US $0.60 per litre of diesel equivalent in Europe (based on rapeseed) and about US$0.50/litre in the United States (based on soybean). Although there exists some scope for cost reduction through standardization processes and building bigger plants, technical breakthroughs on the standard transesterification process, leading to substantial cost reductions in the future, are not expected. Production costs are projected to fall to just over US $0.30/litre in the United States and US $0.40/litre in Europe in 2030.

The current production cost of FT diesel from biomass is about US $0.90 per litre, based on a woody biomass feedstock price of US $3.6/GJ. This cost is expected to decline to US $0.70 to US $ 0.80/litre in the long term.

2.1.5.4 R&D Challenges

2.1.5.4.1 Ethanol

Significant technological challenges exist for the production of ethanol from woody feedstocks because all the steps of the production process need to be optimised. The areas for further research in production of ethanol from lignocellulosic crops includes pre-treatment of the substrate; identification of specific enzymatic saccharification of hemicellulose for enzymatic hydrolysis; development of organisms that can tolerate the inhibitory compounds generated during pre-treatment at the sugars fermentation stage; and product separation techniques. Another research area is to explore possibility of producing all required enzymes within the reactor vessel, thus using the same "microbial community" to produce both the enzymes that break down cellulose into sugars and those that ferment the sugars to ethanol. This "consolidated bioprocessing" is seen by many as the logical end-point in the evolution of biomass-conversion technology.

2.1.5.4.2 Biodiesel

The major challenges include development of more efficient trans-esterification processes, developing new, value-added uses for glycerol and other co-products, biodiesel standardization for quality and performance and also storage stability. Other related research area is minimizing economic cost of bio-diesel production from non-edible oils in degraded and/or wastelands. Another major research area is to perfect the biomass-to-liquid technologies that makes biofuel a competitive option.

2.1.6 Wind Energy

2.1.6.1 Technology

Wind power plants convert the kinetic energy contained in moving air into mechanical rotation energy and subsequently into electricity. Theoretically, a maximum of just under 60 per cent of the power contained in the wind can be extracted. The commercial and technological development of wind energy has been closely related to turbine size. From 10 metres in the mid-1970s, wind turbines have grown to diameters of 126 metres, with multi-MW installed power. Modern wind turbines are designed to have a lifetime of twenty years. Other technological developments include variable-pitch rotors, direct drives, variable-speed conversion systems, power electronics, better materials and improved ratios between the weight of materials and generating capacity.

There are important economies of scale to be achieved in wind turbines. Larger machines can usually deliver electricity at a lower average cost than smaller ones. The reason is that the cost of foundations, road building, maintenance, electrical grid connections and a number of components in the turbine are largely independent of the size of the machine. Large turbines with tall towers use wind resources more efficiently.

2.1.6.2 Cost Projection

Costs of wind power installations depend on system components and size as well as on the site while generating capacity is primarily determined by the rotor-swept area and local wind patterns. The cost of wind power has been declining steadily and the costs at the very best sites have been achieved to about US$ 0.03 to US$ 0.04 per kWh (excluding cost on grid integration and cost attributed to intermittency). For offshore installations, the foundation accounts for one-third or more of the cost. As per IEA estimates annual operating costs are in the range of 1 per cent of the investment cost in the first year to around 4.5 per cent after 15 years. Maintenance and repair costs account for roughly one-third of total operating costs.

Wind energy is expected to reduce at the rate of around 2-3 per cent per annum. However, the cost of electricity will be largely site specific with lower generation cost at high wind speed sites. Deep-water wind costs are projected to decrease to nearly the same level as shallow water costs by 2015 and to reach US $ 0.04 per kWh by 2025. As such with government support for its development, wind power is likely to become generally competitive with conventional technologies between 2015 and 2020.

2.1.6.3 R & D Focus

Increased penetration of wind power will require improved site assessment as a 10 per cent increase in wind speed will result in an energy gain of 33 per cent. Harnessing of offshore wind energy potential on competitive basis will require new approaches in foundation technology together with multi-megawatt-sized wind turbines. The research priorities include increasing the accuracy of forecasting power performance, reducing uncertainties related to engineering integrity, improvement and validation of standards, reducing the cost of new storage options, expanding the cost effective use of existing storage such as hydropower dams, enabling large-scale use and minimising environmental effects. In addition, research focus will be on new design features, such as tall towers made of lightweight materials and advanced aerofoils and on further advances in power electronics. The feasibility of floating wind turbines, individually and in multi-unit formations, will also be in focus. Further, offshore wind development and the role of wind energy within hydrogen-based energy systems will reamain R&D priority areas for the long term. Research needs for wind energy have been summed up in Table 2.7.[9]

Table 2.7 Major R&D Areas in Wind Energy

Area	*Activity*
Cost reduction	Improved site assessment.
	Identification of new high wind sites.
	Better models for aerodynamics and aero-elasticity.
	New intelligent structures/materials and recycling.
	More efficient generators and converters.
Minimizing uncertainties	Forecasting power performance.
	Engineering integrity, improvement of standards.
	Assessment of storage techniques such as pumped hydroelectric storage and compressed air energy storage.
	Enhanced battery based storage options for small turbines.

2.1.7 Hydrogen

2.1.7.1 Description

Hydrogen is a carbon-free energy carrier that has many potential applications. It can be used in modified boilers, engines and turbines, or in fuel cells. The main advantages of using hydrogen as a fuel include zero/no emission of local air pollutants, zero greenhouse gases emissions during its conversion, provided that it is produced from a non-carbon energy source. However, hydrogen is not a primary source of energy and it can be produced from fossil fuels or from nuclear or renewable energy by a number of processes. These include water electrolysis, natural gas reforming, gasification of coal and biomass, water-splitting by high-temperature heat or thermo chemical processes, photo electrolysis, and biological processes. The production cost depends on feedstock and the process adopted. Electrolysis is comparatively costly process. Hydrogen production based on coal gasification and the water-gas shift reaction is also an established technology, the cost of which is higher than production from natural gas. Hydrogen production from high-temperature processes based on nuclear or solar energy is another option, however, they are still far from being commercially viable. The projected costs of gaseous hydrogen in $/GJ are as given in Table 2.8.[10]

Table 2.8 Projected Costs of Hydrogen

	Near-term	*Long-term*
Renewable sources		
Hydrogen from biomass gasification in large plant (18,000GJ/day)—Technology still remains to be demonstrated on a commercial scale. Assumes fuel cost in range $2-4/GJ		7-10
Electrolytic Hydrogen (180 GJ/day)		
Solar PV	24-41	15-25
Wind	20-45	17-25
Solar thermal	45-75	25-35
Fossil Sources		
Steam Reforming natural gas assumes a gas price of $2.5/GJ for large plants and $4/GJ for small plants. 45-60% projected H2 costs due to natural gas.		
• Large plant (144 tonnes/day)	4-7	4-7
• Small plant (1.4 tonnes/day)	11-14	11-14
Coal gasification (Assumes coal prices at $1.5/GJ)		
• Large plant (18,000 GJ/day)	9	
• Medium plant (9,000 GJ/day)	13	

2.1.7.2 Uses

Hydrogen can be used as a combustible fuel in internal combustion engines or turbines. The efficiency of hydrogen combustion in engines is comparable to that of petrol with state-of-the-art technology. Both in stationary and mobile applications, use of hydrogen reduces the problems of the local emission of pollutants, even when 'conventional' technologies are used, as the exhaust gas does not contain carbon monoxide, sulphur dioxide, hydrocarbons, lead compounds, soot particles, etc. Specific control has to be done in order to manage NOx emissions.

Hydrogen can also be used with other technologies that are less suitable for hydrocarbons and require hydrocarbons to be reformed before hand. The best known process is fuel cell technology. Fuel cells use hydrogen or a hydrogen-rich fuel such as natural gas and oxygen to produce electricity through an electrochemical process. A fuel cell consists of two electrodes—a positive—anode and a negative—cathode—sandwiched around an electrolyte. They operate by feeding hydrogen to the anode and oxygen to the cathode. Activated by a catalyst, the hydrogen atoms separate into protons and electrons, which take different paths to the cathode. The electrons go through an external circuit creating electricity while protons migrate through the electrolyte to the cathode, where they reunite with oxygen and electrons to produce water and heat. Fuel cells are highly efficient, clean and quiet, with no moving parts. Fuel cells are available as pilot and demonstration systems and in small batches for a wide range of outputs from a few watts to units with several kW. They run at temperatures of 80–800°C.

There are five main classes of fuel cell, each with differing characteristics and operating temperatures: the alkaline fuel cell (60-90°C), the solid polymer fuel cell (80-100°C), the phosphoric acid fuel cell (200°C), the molten carbonate fuel cell (650°C), and the solid oxide fuel cell (800-1000°C). The low temperature fuel cells generally incorporate precious metal electrocatalysts, exhibit fast response and short start-up times but require a relatively pure supply of hydrogen as catalysts can be poisoned by carbon monoxide. They are available commercially or are near commercialisation. The higher temperature fuel cells, in contrast, can be directly operated on a range of light hydrocarbon fuels and do not require expensive electro-catalysts, and generate useful heat. However, they have long start-up times, reliability is still a concern and are only at the demonstration stage.

2.1.7.3 Cost Reduction

Current fuel cells cost significantly more than the base line options for both automotive and stationary applications. Among all fuel cell technologies, proton exchange membrane (PEM) fuel cells are particularly suited to powering passenger cars and buses, and have now been installed in demonstration vehicles. PEM fuel cell drive systems cost more than US$ 2 000/kW. Estimates indicate that if fuel cell cars were mass-produced, the cost of fuel cell drive systems could be reduced to US$ 100/kW. This implies that the costs of the fuel cell stacks must fall below US$ 50/kW to make them competitive with internal combustion engines. The investment costs for stationary fuel cells are between US$ 13 000 and US$ 15 000 per kW, compared to US$ 500 to US$ 700 per kW for a conventional large-scale gas-fired combined-cycle power plant. With mass production, the cost of molten carbonate fuel cell systems are expected to reach a target level of about US$ 1650 per kW.

2.1.7.4 R& D Focus

In fuel cell technologies the research direction will be towards cost reduction, improving durability and standardization of processes. Areas for cost reduction will include development of newer, cheaper catalysts and stable membrane. R&D is needed to reduce costs and increase the efficiency of hydrogen production and the production of hydrogen from photo-electrolysis and from biological production processes are both at a very early stage of development and would require concerted efforts to reach competitive levels. Hydrogen production through process such as thermo-electrochemical and electrochemical process are at advance stages of R&D. R&D on balance of plant also has to be given due importance for improving system integration and reliability of the fuel cell system in whatever application it is used in. Removal of market barriers and development of hydrogen infrastructure facilities is equally important to lead fuel cell technology into commercialisation.

NOTES

1. Adopted from IEA; Energy Technologies Perspective: Scenarios and Strategies 2050.
2. Gregory F. Nemet, Technical Change in Photovoltaics and the Applicability of the Learning Curve Mode, Interim Report IR-05-029. International Institute for Applied Systems Analysis, Lxumburg.
3. Green, M A, Emery, K, King, D L, Igari, S and Warta, W (2002) Progress in Photovoltaics. *Research and Applications* **10**: pp 55–61, UNDP (2000) *World Energy Assessment. Energy and the Challenge of Sustainability*. UNDP, New York. As referred in German Advisory Council on Global Climate Change (WBGU) Report, 2004.
4. Adopted from German Advisory Council on Global Climate Change (WBGU) Report, 2004.
5. Adopted from German Advisory Council on Global Climate Change (WBGU) Report, 2004.
6. World Energy Council, 2000.
7. International Energy Agency: World Energy Outlook 2006.
8. International Energy Agency, World Energy Outlook 2006.
9. Adopted from Energy Technologies Perspective, IEA 2006.
10. Adopted from Imperial Colleges's ICCEPT Report 'Assessment of Technological Options to Address Climate Change'.

Chapter 3

Renewable Energy Policies: EU, Germany, USA, Japan and China

Rapid price increases on the fossil fuel markets in 1970s, was the main driver for many countries to develop renewable energy technologies for meeting domestic energy needs. Over the years development of renewable energy was not only pursued as a way to reduce dependence on imports of fossil fuels but also for environmental considerations. In developing countries context renewable energies were not only viewed as promise to meet energy needs domestically but as the only technological solution to provide access to energy for millions living in rural and remote areas. Since then significant development has taken place and many of the renewable energy technologies have crossed the demonstration phase and are being used for large scale applications.

In 2005, renewables accounted for around 13 per cent of the world total primary energy supply. Combustible renewables and waste (97 per cent of which is biomass, both commercial and non-commercial) represented 79.4 per cent of total renewables followed by hydro (16.7 per cent). Renewables accounted for over 18 per cent of electricity production in 2005, after coal (40 per cent) and natural gas (close to 20 per cent), but ahead of nuclear (16 per cent), and oil (7 per cent) and non-renewable waste. Almost 90 per cent of electricity generated from renewables comes from hydropower plants while close to 6 per cent comes from combustible renewables and waste. Geothermal, solar and wind have now reached 4.5 per cent of renewable generation.[1]

Although a large number of countries have encouraged application of renewable energy technologies, the regional group and countries implementing fairly large and ambitious programmes are EU, Germany, USA, Japan, China and India. In order to understand the process, renewable energy policies of these countries and EU have been analyzed.

The country choices provide a mix of different conditions in terms of availability of resources, geo-political systems and traditions not only inter-country but also within the countries. A major policy that will directly affect the energy choices of EU countries and Japan is emission reduction targets under the Kyoto Protocol. Although the United States is not a signatory to the Kyoto Protocol it has undertaken proactive policies that addresses climate change concerns and has been attempting to address climate change through technology development paths including IPHE (International Partnership for Hydrogen Energy) and Asia Pacific Climate Change Partnership. China and India have no commitments for emission reduction and access to energy in an affordable and sustainable manner is main driver for renewable energy.

3.1 EU Renewable Energy Policy Framework

The population of the European Union represents more than 7 per cent of the world's population but produces more than 25 per cent of the world GDP (Gross Domestic Product), thus making EU-25 the second largest world economy. EU-25 consumes more than around 16 per cent of the world's total primary energy supply (TPES) and emits around 15 per cent of the total CO_2 emissions. EU has been continuously evolving in terms of its framework on various fields of common interest and also number of member countries. There are 25 EU member countries[2], however, before 1 May 2004 EU consisted of 15 member states. EU member countries work together, with administrative support from the European Commission (EC), which reports to an elected European Parliament. EC is responsible for proposing policies and legislation and once adopted at the EU level, each Member State is required to implement policies and translate the EU Directives into national legislation as per national circumstances.

Climate change concern is one of the main driving forces for moving towards renewable energy technologies and to this effect EU member countries have initiated strong policy measures towards increasing share of renewable energy in their energy mix and move towards decarbonising their economy. Under the Kyoto Protocol to the United Nations Framework Convention on Climate Change, EU has to reduce its green house gases emission by 8 per cent to its emission level in 1990 by 2012.[3] EU member states further divided the overall emission reduction target among themselves. Consequently the member states initiated specific measures to achieve their emission reduction targets. The council of the European Union passed a directive for establishing a scheme for greenhouse gas emission allowance trading within the Community and which took effect on 25 October 2003.[4] Under the directive each Member State has to establish a national allocation plan for each trading period, with the first 2005-2007. In this allocation the Member State decides a total number of allowances to be created for the period and the distribution of these allowances to individual plants. An installation that emits more CO_2 than it has allowances for would need to buy additional allowances in the market, while one that emits less has the possibility to sell its surplus allowances. Under the EU Emissions Trading Scheme (EU-ETS), some 12,000 energy-intensive plants across the EU are able to buy and sell permits to emit carbon dioxide, covering about 40 per cent of the EU's total CO_2 emissions. Industries covered by the scheme include: power generation, iron & steel, glass, cement, pottery and bricks. An emission cap is defined, for each individual plant, in a National Allocation Plan (NAP) submitted by member states and approved by the Commission. Companies exceeding their quotas are allowed to buy unused credits from those doing better at cutting their emissions. A fine of euro 40 per excess tonne of CO_2 emitted will be imposed on plants exceeding their individual target, rising to euro 100 in 2008. For comparison, carbon prices have fluctuated between euro 8-30 a tonne in 2005-06 (one tonne = one allowance). By offering a much cheaper alternative to fines, the Commission hopes that the EU-ETS will stimulate innovation and create incentives for companies to reduce carbon emissions.[5] The implementation timeline are 2005-2007 as first trading period and 2008-2012, the second trading period, coinciding with period under which Kyoto commitments are to be achieved.

EC is promoting the deployment of renewable energy technologies across the EU to address three main energy policy challenges, viz emissions reduction targets under the Kyoto Protocol to the UNFCCC; high, and steadily rising, level of European energy import dependence; and potential of renewable energy sources as a valuable instrument for economic development and greater social cohesion within the EU.

The EU started working towards a policy framework for renewable energy in the early 1990s, and a White Paper and Action Plan for Renewable Energy Sources was published in 1997.[6] The main component of the White Paper was setting the EU the target of increasing the share of renewable energy to 12 per cent of total energy consumption by 2010. It also detailed targets for each renewable energy technology.

A step towards achieving the objectives of the white paper was the Directive 2001/77/EC on the promotion of electricity produced from renewable energy sources in internal electricity market.[7] The directives came into effect on 27 October 2001. These directives aim to increase renewable energy's share to 12 per cent of primary energy consumption in the EU and to boost renewable sources to 22 per cent of electric power production by 2010. Within the directive, indicative targets are set for each member state, whereby the respective country is free to decide which instruments to use to reach the indicative targets. EC in its 2005 report, has quantified member—country wise progress in achieving the aim set through 2001 directives.[8]

3.1.1 Research and Development

Joint development programmes have been part of the European Union virtually since its conception in the 1950s. EC has been supporting research and technological development in renewable energy since 1970s and it still continues to be high priority. Since 1984, research and innovation activities of the EU are grouped in one programme called the Framework Programme (FP). Research framework programmes are the EU's main financial and legal instruments to implement the European Research Area (ERA). As such in 1984 the Commission launched the 1st five-year Framework Programme; a second launched in 1987; a third in 1990; a fourth in 1994; a fifth in 1998; and a sixth in 2002. The EU's Framework programmes cover all non-nuclear energy R&D financing for research, development and demonstration projects across the whole energy sector. This includes funds for energy efficiency, renewable energy and cleaner fossil fuels. The total budget allocated for energy in 4th, 5th and 6th Frameworks were Euro 2412 million, Euro 2302 million and Euro 2040 million respectively. Of the total allocation non-nuclear research support was around 40 per cent.

The Framework Programmes for Research and Development (FP) of the European Union have contributed from their beginning to the development of renewable energy technologies. These Community actions have a proven European added value in terms of building critical mass, strengthening excellence and exercising a catalytic effect on national activities. In combination with national activities, working at European level with an adequate combination of innovation and regulatory measures has produced substantial results.

The non-nuclear research and technology development portfolio under sixth framework programme comprised five clusters of technologies, (a) Fuel Cells (FC), including their applications; (b) New technologies for energy carriers/transport and storage, in particular hydrogen; (c) New advanced concepts in renewable energy technologies; (d) Capture and sequestration of CO_2 associated with cleaner fossil fuel plants; (e) Socio-economic tools and concepts for energy strategy.

Table 3.1 European Commission Funding of Non-nuclear Research and Technology Development

Sustainable Energy Systems	Technology Parts (Strategically important areas and topics)	EC Funding in FP5: Number of Projects	FP5: Eligible Costs (Euro million)	FP5: Total EC contribution (Euro million)	FP6: Number of Projects	FP6: Eligible Costs (Euro million)	FP6: Total EC contribution (Euro million)
New	PV	85	268.26	105.30	19	137.11	75.73
Advanced	Bioenergy	93	549.85	110.48	30	250.97	127.36
Concepts in	Wind	20	44.69	24.36	10	79.74	31.59
Renewable	Geothermal Systems	1	24.60	6.50	5	39.49	13.36
Energy	Ocean	7	11.67	6.85	7	30.78	14.03
Technologies	Concentrated Solar Thermal	7	25.10	11.79	6	17.35	10.33
	Total renewables	213	924.17	258.78	77	555.44	272.40
Others	Fuel Cells	41	228.52	97.43	33	286.65	153.92
	CO_2 Storage and capture	9	31.8	16.00	18	121.86	68.71
	Hydrogen	25	72.1	38.57	38	213.99	125.69
	Grid	48	121.06	62.61	15	84.76	50.54
	Socio-economic	11	8.10	5.99	20	23.57	23.59
	Total Others	134	461.58	220.59	124	730.83	422.45
Non-nuclear	Total	347	1,385.75	479.37	201	1,286.28	694.85

The budget for non-nuclear was broadly divided into a share of 40 per cent for renewables, 55 per cent for the other technologies and 5 per cent for coordinating and cross-cutting activities. A summary of funding for various activities is given in the table 3.1.[9]

The seventh Framework Programme (FP7) is for 7-year period from 2007-2013 with total budget of Euro 64.3 billion. Energy research is one of the thematic research area with a budget of Euro 2.3 billion for supporting co-operation between universities, industry, research centres and public authorities across the EU and with the rest of the world.

3.1.2 Regulatory Measures and Financial Incentives

The support for renewable energy sources differs significantly among the different EU Member States, which is mainly due to the different country-specific resource conditions and differences in the support instruments applied in these countries. The country-wise policy overview for renewables is summarized at Table 3.2.[10]

As per its mandate under renewable directives 2001, EC will make an assessment on level of Member States systems for promoting renewables electricity by December 2007 and on the basis on results propose a policy framework for renewable energy beyond 2010 and also targets for 2020.

The European Directive on the promotion of cogeneration calls for harmonization of combined heat and power (CHP) power calculation methods, simplification of grid access, and regular national and European reports. This directive emphasises on accommodating specific national frameworks.[11]

The European directive on the promotion of the use of biofuels or other renewable fuels for transport stipulates that the EU member states should ensure that a minimum share of biofuels and other renewable fuels be introduced in respective markets and by 31 December 2010 this share should increase to 5.75 per cent.[12]

Table 3.2 EU Member Country Renewable Energy Policy

Country	*Main electricity support schemes*	*Comments*
Austria	Feed-in tariffs (now terminated) combined with regional investment incentives.	Feed-in tariffs have been guaranteed for 13 years. The instrument was only effective for new installations with permission until December 2004. The active period of the system has not been extended nor has the instrument been replaced by an alternative one.
Belgium	Quota obligation system/Tradable Green Certificate (TGC) combined with minimum prices for electricity from RES.	The Federal government has set minimum prices for electricity from RES. Flanders and Wallonia have introduced a quota obligation system (based on TGCs) with the obligation on electricity suppliers. In Brussels no support scheme has been implemented yet. Wind offshore is supported at federal level.
Denmark	Premium feed-in tariffs (environmental adder) and tender schemes for wind offshore.	Settlement prices are valid for 10 years. The tariff level is generally rather low compared to the previously high feed-in tariffs.
Finland	Energy tax exemption combined with investment incentives.	Tax refund and investment incentives of up to 40 per cent for wind, and up to 30 per cent for electricity generation from other RES.
France	Feed-in tariffs in place.	For power plants < 12 MW feed-in tariffs are guaranteed for 15 years or 20 years (hydro and PV). For power plants > 12 MW a tendering scheme is in place.
Germany	Feed-in tariffs.	Feed-in tariffs are guaranteed for 20 years (Renewable Energy Act). Furthermore soft loans and tax incentives are available.
Greece	Feed-in tariffs combined with investment incentives.	Feed-in tariffs are guaranteed for 10 years. Investment incentives up to 40%.
Ireland	Tendering scheme. It has been announced that the tendering scheme will be replaced by a feed-in tariff scheme.	Tendering schemes with technology bands and price caps. Also tax incentives for investment in electricity from RES.

(*Contd.*)

Country	*Main electricity support schemes*	*Comments*
Italy	Quota obligation system/TGC. A new feed-in tariff system for photovoltaic valid since 5th August 2005.	Obligation (based on TGCs) on electricity suppliers. Certificates are only issued for new RES-E capacity during the first eight years of operation.
Luxembourg	Feed-in tariffs.	Feed-in tariffs guaranteed for 10 years (for PV for 20 years). Investment incentives also available.
Netherlands	Feed-in tariffs.	Feed-in tariffs guaranteed for 10 years. Fiscal incentives for investment in RES are available. The energy tax exemption on electricity from RES ended on 1 January 2005
Portugal	Feed-in tariffs combined with investment incentives.	Investment incentives up to 40 per cent.
Spain	Feed-in tariffs.	Electricity producers can choose between a fixed feed-in tariff or a premium on top of the conventional electricity price, both are available over the entire lifetime of a RES power plant. Soft loans, tax incentives and regional investment incentives are available.
Sweden	Quota obligation system/TGC.	Obligation (based on TGCs) on electricity consumers. For wind energy, investment incentives and a small environmental bonus are available.
UK	Quota obligation system/TGC.	Obligation (based on TGCs) on electricity suppliers. Electricity companies which do not comply with the obligation have to pay a buyout penalty. A tax exemption for electricity generated from RES is available (Levy Exemption Certificates which give exemption from the Climate Change Levy).

(*Contd.*)

Country	*Main electricity support schemes*	*Comments*
Cyprus	Grant scheme for the promotion of RES (since February 2004) financed through an electricity consumption tax of 0.22 E/kWh (since Aug. 2003).	Promotion scheme is fixed only for a 3-year period.
Czech Republic	Feed-in tariffs (since 2002), supported by investment grants Revision and improvement of the tariffs in February 2005.	Relatively high feed-in tariffs with 15-year guaranteed support. Producer can choose between a fixed feed-in tariff and a premium tariff (green bonus). For biomass cogeneration, only the green bonus applies.
Estonia	Feed-in tariff system with purchase obligation.	Feed-in tariffs paid for up to 7 years for biomass and hydro and up to 12 years for wind and other technologies. All support schemes are scheduled to end in 2015. Together with relatively low feed-in tariffs this makes renewable investments very difficult.
Hungary	Feed-in tariff (since January 2003) combined with purchase obligation and tenders for grants.	Medium tariffs (6 to 6.8 ct/kWh) but no differentiation among technologies. Actions to support RES are not coordinated, and political support varies. All this results in high investment risks and low penetration.
Latvia	Quota obligation system (since 2002) combined with feed-in tariffs.	Frequent policy changes and the short duration of guaranteed feed-in tariffs result in high investment uncertainty. The high feed-in tariff scheme for wind and small hydropower plants (less than 2 MW) was phased out in January 2003.

(Contd.)

Country	Main electricity support schemes	Comments
Lithuania	Relatively high feed-in tariffs combined with a purchase obligation. In addition good conditions for grid connections and investment programmes.	Closure of the Ignalina nuclear plant will strongly affect electricity prices and thus the competitive position of renewables as well as renewable support. Investment programmes limited to companies registered in Lithuania.
Malta	Low VAT rate for solar.	Very little attention to RES-E so far.
Poland	Green power purchase obligation with targets specified until 2010. In addition renewables are exempted from the (small) excise tax.	No penalties defined and lack of target enforcement.
Slovakia	Programme supporting RES and energy efficiency, including feed-in tariffs and tax incentives.	Very little support for renewables. The main support programme runs from 2000, but there is no certainty as to the time frame or tariffs. The low support, lack of funding and lack of longer-term certainty make investors very reluctant.
Slovenia	Feed-in system combined with long-term guaranteed contracts, CO_2 taxation and public funds for environmental investments.	None.

Further the directive on restructuring of community framework for the taxation of energy products and electricity grants member states a mineral tax reduction, extending to a complete exemption for pure biofuels, or the biogenic component of fuel.[13]

The European Directive on the energy performance buildings is to support the improvement of the energy performance of buildings in consideration of the respective external climatic and local conditions, the interior climate requirements, and cost-effectiveness. For new buildings decentralized power supply system based on renewable energy sources constitutes one of the factors to be considered.[14]

In order to give further impetus to the aims set under the European Energy policy, EU launched 'Campaign Sustainable Energy 2005-2008' in 2005. Success of the campaign was set to be measured by the indicative targets as given in Table 3.3.[15]

Renewable energy sources are likely to remain a focal point of discussion for the future EU energy policy as they contribute to overall aims of European energy policy, viz, (a) security of energy supply; (b) competitive energy costs & prices; and (c) environmentally benign cost of energy. With the high price levels on fossil fuel markets, increasing import dependence coupled with climate change concerns, will place the use of renewable energies on the agenda in a medium—and long-term energy policy strategy.

Germany has one of the most vibrant renewable energy programmes among EU Member States. The share of renewable electricity in Germany has more than doubled within the past decade. While Germany is still far from a sustainable electricity supply, relying heavily on coal (50 per cent) and nuclear (28 per cent), no other country has been successful in growing new capacity as quickly as Germany, particularly in the wind power sector. Germany accounts for over 40 per cent of installed world wind power capacity. The country is an interesting success story for renewable energy development. Therefore, Germany has been chosen as a leading EU member state for understanding the drivers and dynamics behind this successful growth.

Table 3.3 Achievable Benchmarks for 2008 Arising from EU Policy and Legislation

Renewable Energy Sources	*Indicative target for renewable electricity have been set by the European Union at 22.1 per cent of total electricity production by 2010*
Wind	15,000 MW new capacities of windturbine
Solar thermal	35 million m_2 of solar thermal installations
Photovoltaic	1500 MWp of photovoltaic installations
Geothermal	15 new power plants and 10 new low-mid temperature plants and 250,000 new installed geothermal heat pumps
Small hydro	2,000 MW new capacities of small hydro installations
Biogas	6,000 new biogas plants
Biomass	450 new combined heat and power plants and 13,000 new district/centralized heating unit installations
Fuels for transport	Indicative target for biofuels have been set by European Union at 5.75 per cent of total fuel for transport use by 2010
Bioethanol	Fivefold increase in the production of bioethanol
Biodiesel	Threefold increase in the production of biodiesel
Energy performance in buildings	The EU Directive on the Energy Performance of Buildings points to energy saving potential of 22 per cent in building sector by 2010
General	2 million inspections and assessments of cooling systems 10,000 new Energy services contracts between public authorities and private companies
Existing residences (small buildings)	5 million inspections and assessments of heating systems 2 million new Energy performance certifications
Existing residences (single and multi-family housing)	10 million residences to reduce energy consumption by 30-40 per cent compared to actual consumption 1 million additional residences with 50 per cent of delivered energy provided by renewables
Existing buildings (> 1000 m_2)	100,000 Energy performance certifications 50,000 additional buildings with 50 per cent of delivered energy provided by renewables
New residences (single family housing)	50,000 'very low' energy houses built
All residences in EU	1 low-energy appliance and 1 low-energy light source (CFL)—monitoring at the level of sales

3.2 Renewable Energy in Germany

The Federal Republic of Germany has a population of 82 million and is one of the largest economies in the world. It has a total surface area of 35.7 million hectares, comprising farmland (53.5 percent), woodland (29.4 percent), settlements (12.3 percent) and other areas (4.7 percent).[16] Due to size and structure of its economy, Germany is one of the largest carbon emitters in European Union. It is responsible for approximately 800 million tonnes of carbon dioxide (CO_2) emission annually, accounting for about one-fourth of EU greenhouse gas emissions. Within the burden sharing agreement under the Kyoto Protocol, Germany is committed to reduce carbon emissions by 21 per cent in 2008-2012 compared to 1990. A substantial portion of greenhouse gas emissions is produced by the electricity system, as CO_2 emissions due to fossil fuel combustion for electricity production amount to more than 40 per cent of total CO_2 emission in Germany.[17] At the same time, Germany is facing a major renewable and restructuring process. Around one-third of its total electric generating capacity, in the form of fossil fuel based generation, may retire within next 20 years and another one-sixth in the form of nuclear power plants[18], is scheduled to be phased out. With projected stable electricity demand, this means that almost 50 per cent of German electric power capacity could be replaced within next 20 years.[19] This as such provides substantial window of opportunity for new and innovative technologies including renewables.

3.2.1 Achievements

The share of renewable power in installed non-hydro renewable electricity generation capacity in the erstwhile West Germany was very small until the early 1990s, when installation of wind turbines and solar panels began to increase. The share of renewables in the total energy and electricity mix has been continuously growing since 1990s, when Germany initiated specific policy measures for harnessing these sources. By 2005, the share of renewable energy sources was 6.4 per cent to the final energy consumption. These sources contributed 10.2 per cent of the gross electricity consumption, 5.3 per cent of heat allocation and 3.6 per cent of fuel consumption.[20]

Table 3.4 Share of Renewable Energy[21]

		1990	*2000*	*2005*
Hydro power	GWh	17000	24936	21524
	MW	4403	4572	4680
Wind Energy	GWh	40	7550	26500
	MW	56	6112	18428
Biomass Electricity	GWh	222	2279	11394
	MW	190	664	2192
Biogenic share of waste Electricity	GWh	1200	1850	2050
Photovoltaics	GWh	1	64	1000
	MW	2	100	1508
Geothermal electricity	GWh	0	0	0.2
	MW	0	0	0.2
Total Electricity generation	GWh	18463	36679	62468
Share of gross electricity	%	3.4	6.3	10.2
Biodiesel	GWh	N/A	2583	18600
Vegetable oil	GWh	0	0	2047
Bioethanol	GWh	0	0	1683
Total biofuels[22]	GWh	0	2583	22330
Share of primary energy consumption	%	N/A	2.6	4.6

Figure 3.1 depicts the growth of renewable energy shares in Germany between 2000 and 2005, respectively for Total Primary Energy Consumption, Gross Electricity Consumption, Final Energy Consumption for heat, fuel for road traffic and Final Energy Consumption.

In 2005, Germany generated about 10.2 per cent of electricity from renewable sources, with non-hydro renewable sources contributing 6.7 per cent. Apparently, this is attributed to strong growth in wind energy in which Germany has emerged as the world leader and by the end of 2005, 17,574 wind turbines with a total capacity of 18,428 MW were in operation in Germany.[23]

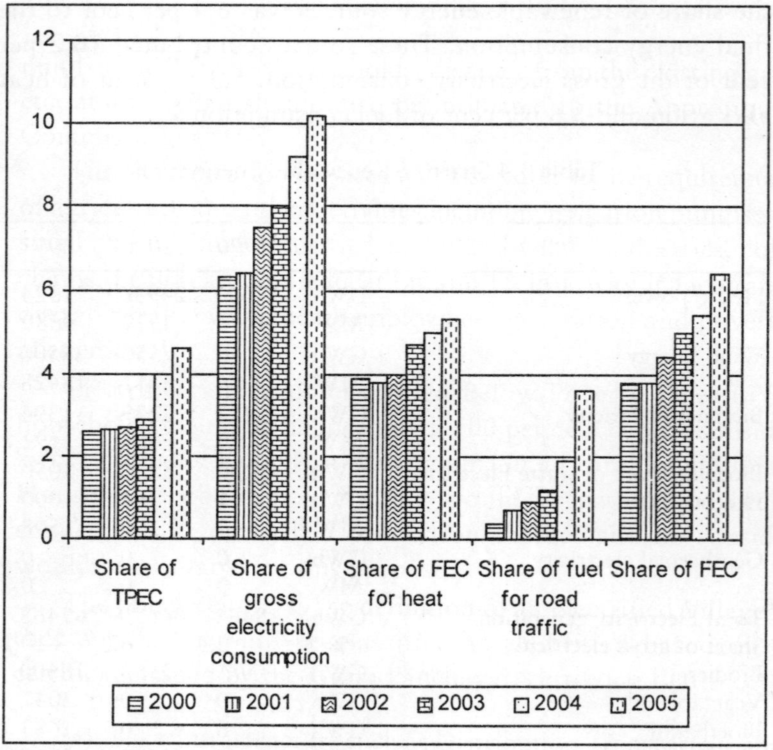

TPEC: Total Primary Energy Consumption; FEC: Final Energy Consumption

Figure 3.1 Growth of Renewable Energy Share in Germany

In 1998 the German Federal Government had set a goal to increase renewable energy sources' (RES) share in overall energy supply to at least 4.2 per cent by 2010, which has already been achieved in 2005. The target for 2010 is that renewable energies' share in gross electricity consumption should have risen to at least 12.5 per cent. In addition, a medium-term goal in the Renewable Energy Sources Act of increasing renewable energies' share in the electricity supply to at least 20 per cent by 2020 has been set. This apart, the aim is to increase renewable energies' share in overall energy consumption to at least 10 per cent and biofuels' share in transport fuels to 5.75 per cent by 2020.[24]

Over the past two decades, Germany's approach towards renewable energy has shifted from a focus on publicly-financed R&D towards policies promoting application and implementation of new technologies in the market place. This transition has occurred gradually and as a result of a number of domestic and international political developments since the 1980s. EU Directives mandating aggressive renewable energy deployment, pressure from domestic environmental advocates and legislators, legislation to the phase-out of nuclear power by 2020, and commitments to international treaties such as the Kyoto Protocol, all have been important factors in the evolution of German energy policy. The deployment-based policies occurred implicitly and incrementally, in response to political constraints and limited energy technology options, rather than through conscientious changes in the direction of strategic policy or a belief that demand-increasing policies are fundamental drivers of technological advance.

3.2.2 Research and Development

Since 1974 to around late 80's the renewable energy policy in Germany was focused on promotion of research, from training personnel to development of prototypes and laboratory production. In 1974 around euro 10 million were spent on these activities which rose gradually until 1978 to about euro 60 million and in 1982 the total spending on renewable energy was around euro 150 million. The spending on renewables thereafter declined and in 1986 the spending was euro 82 million only.[25] The German Government has spent almost euro 2 billion on these activities over more than three decades and research is still an important part of renewable energy programmes. In 2005, the Federal Environment Ministry (BMU) approved a total of more than 100 new projects on photovoltaics, wind power, geothermal power, low-temperature solar thermal heating, solar thermal power stations, horizontal topics and other issues with a total financial volume of around 98 million euros. As summary, in 2005 in all

481 projects with total funding of euro 293 million were ongoing and during the year 192 projects with total funding of euro 90 million were completed.[26] It may be mentioned that in addition to BMU, Ministry for Economy and Technology (BMWi) and Ministry of Agriculture have also been supporting research in the fields of hydrogen & fuel cell and bio-energy. Renewable energy research in Germany is being pursued under its 5th Energy Research Programme (2003-2008) with overall funding for renewables to the order of around euro 750 million. The earmarked funding for 2006-2008 period is euro 412.9 million. Specific institutions have been entrusted with the responsibility for managing the research area and these institutions submit analysed project with their recommendation for funding or otherwise.[27]

3.2.3 Wind Power

The wind power sector is a particularly good example for technological change and the emergence of a renewable energy industry in Germany. The development of wind turbines had two very different roots.[28] The Große Windenergieanlage Bau-und-Betriebsgesellschaft mbH (GROWIAN) project in the late 1970s and early 1980s was a top-down approach by government and established research and industry players aimed at building a large (3MW) wind turbine from scratch. This eventually failed. On the other hand, a more successful approach to wind turbine development, although much less visible in the beginning, has been pursued by several small new entrants entering in the mid-1980s.[29] The size of newly installed turbines increased from 10 to 50 kW in the 1980s to an average of 182 kW in 1992. The first wind energy programme in Germany was the "100 MW Programme," launched in 1989 by the Ministry for Research & Technology. Though the programme was initially intended as a market introduction mechanism for wind energy, it was determined that the Ministry for Research could not legally sponsor the program without a research component. Thus, wind turbines in the programme were paid a subsidy of 8 Pfennige

(0.04 euro) per kilowatt-hour and were obliged to report their turbines' performance. The programme proved successful from the outset and was quickly increased to the "250 MW" programme a year after its inception. Though the subsidy was reduced to 6 Pfennige (0.03 euro) in 1991, the program met with success until it ended in 1996 after 350 MW had been installed. The information on turbine performance, and the unintentional research component has turned out to be a valuable resource for developers, planners and manufacturers.[30]

3.2.4 Solar Roof Programme

In 1990, the German government began the "1,000 Roof" program, which sought to move some of the government's renewable energy focus away from R&D and towards demonstration projects. From 1991 to 1995, under the 1,000 roof programme applicants received 50 per cent funding of investment costs from the federal government plus 20 per cent from the land government. Eventually 2.250 roofs were equipped with PV modules, leading to about five MW of installations.[31] To further promote the development of PV power, on 1 January 1999, the German government launched the 100,000 Roof Programme for PV systems, aimed at stimulating the installation of 100,000 grid-connected PV systems totalling 300 MWp within 6 years. In the beginning, the programme granted loans at a "reduced" interest rate of 0 per cent for PV systems and a waiver of the last instalment of up to 12.5 per cent of the investment. In total, this corresponds to a subsidy rate of about 35 per cent. In 2003, the 100,000 Roofs Programme ended successfully, after loans for more than 300 MWp PV systems had been granted. In order to fill the gap created by the termination of the programme, the government decided to increase the feed-in tariff provided by the German Renewable Energy Act. Since January 2004, grid-connected PV systems on roofs and facades receive a tariff between 0.540 euro and 0.624 euro per kWh over 20 years.[32]

3.2.5 Electricity Feed Law

A federal Electricity Feed Law (EFL) adopted in 1991 was the first major initiative for the promotion of renewable energy in Germany. It obligated public utilities to purchase renewably-generated power from wind, solar, hydro, biomass and landfill gas sources, on a yearly fixed rate basis, based on utilities' average revenue per kWh. Remuneration to wind producers was set at 90 per cent of the average retail electricity rate, for other renewable power providers, compensation was set at 65-80 per cent depending on plant size, with smaller plants receiving the higher subsidy level. The EFL effectively subsidized the operation of commercial wind installations at 4.1 Euro cents/kWh, and jump-started wind power's market breakthrough in the 1990s. In addition, investment in wind power installations was also subsidized by a domestic, state-owned development bank, the Deutsche Ausgleichsbank, which offered low-interest, government guaranteed loans for new wind power development.

3.2.6 Renewable Energy Law

The successor to the EFL was the German Renewable Energy Law (Erneuerbare-Energien-Gesetz—EEG), first introduced on 29 March 2000 and last amended on 21 July 2004 ensures the increased use of environmentally friendly renewable energies, not through subsidies but through apportioning the costs. The EEG guarantees operators of renewable energy systems priority for feeding their electricity into the grid and payment at a fixed reimbursement rate over an assured period of time and as such this arrangement provides investment security. The reimbursement payments are apportioned to all electricity consumers through a national equalisation mechanism, which also disencumbers power-intensive companies so that they do not suffer any competitive disadvantages. The degressive reimbursement rates for future systems provide an incentive for innovations and cost reductions. The core elements of the EEG are: (a) the priority connection of installations for the generation of electricity from renewable energies and from mine gas to the general electricity supply grids; (b) the priority purchase and transmission of this electricity and a consistent fee for this electricity paid by the grid operators, generally for a 20 year period, for commissioned installations; (c) the nationwide equalisation of the electricity purchased and the corresponding fees paid; (d) The fee paid for the electricity depends on the energy source and the size of the installation and date of commissioning. As such this law has dictated that the costs of grid connection for renewables projects are the responsibility of the utility, which can pass on the costs to consumers.

The EEG is intended to contribute to the increase in the percentage of renewable energy sources in power supply to at least 12.5 per cent by 2010 and to at least 20 per cent by 2020. Unlike that of the EFL, the EEG's remuneration system is not based on average utility revenue per kWh sold, but rather on a fixed, regressive feed-in tariff for renewable sources. Low-cost renewable energy producers are compensated at lower rates than higher-cost producers, providing strong incentives for the development and operation of renewable energy installations on lower-quality sites. Also, under the EEG, grid operators are obligated to purchase power from local producers. The nationwide equalization scheme has been implemented to reduce the cost differentials paid by grid operators in different parts of the country for the purchase of renewably-generated electricity.[33]

EEG has proven to be an extremely successful instrument for expanding renewable energy in the German electricity market. Not only are the national goals being met, but also the European guidelines for the promotion of electricity generation from renewable energy since 2001. In so far as additional cost burden on the consumers due to EEG provisions is concerned, as per the studies conducted by the German Federal Ministry for the Environment, Nature Conservation and Nuclear Safety such apprehensions do not hold ground. The fee payments to plant operators totalled almost euro 4.1 billion in 2005. The resulting additional cost (differential cost) as compared with the costs for electricity generated from conventional energy forms equalled around euro 2.4 billion.

Table 3.5 Feed-in Rates under the Renewable Energy Sources Act of 2004[34]

Source	*Rate in euro cent/kWh*	*Annual decline*	*Comments*
Methane from landfill, coal mines, sewage stat.			Additional 2 cents if certain innovative technologies are used.
up to 500kW	7.67	1.5 per cent	
up to 5MW	6.65		
above 5 MW	6.65		
Hydro*			In many cases only the additional capacity resulting from moderni-sation of the installation receives the special rate.
up to 500kW	7.67/9.67	1 per cent, begining in 2005	
up to 10MW	6.65		
up to 20MW	6.10		
up to 50MW	4.56		
up to 150MW	3.70		
Biomass			There is a system of bonuses (for innovative solutions, for CHP etc.) which can raise the rate up to 21.5 cents.
up to 150kW	11.50	1.5 per cent, beginning in 2005	
up to 500kW	9.90		
up to 5MW	8.90		
Above 5MW	8.40		
Recycled timber (categ. A III/IV)	3.90		
Geothermal			
up to 5MW	15.00	1 per cent, beginning in 2010	
up to 10MW	14.00		
up to 20MW	8.95		
above 20MW	7.16		
Wind power		2 per cent, beginning in 2005 for onshore and 2008 for off-shore	On-shore: Turbines with a yield of less than 60 per cent of reference turbine are not eligible for tariff. Off-shore: Turbines located in nature or landscape or bird protection areas are not eligible.
On-shore: initial rate (at least five years)**	8.70		
definitive rate	5.50		
Off-shore: initial rate	9.10		
(12 years or longer***) Definitive rate	6.19		
Solar radiation		5 per cent, beginning in 2005. For non-built up areas, the decline increa-ses to 6.5 per cent beginning in 2006	There is a bonus of 5 cents for the systems set up at façade of building.
Attached on or to building			
up to 30kW	57.40		
up to 100kW	54.60		
Above 100 kW	54.00		
Other systems	45.7		

Note:* Except hydro (15-30 years or longer depending upon category) in all other cases the duration of receiving special rate by an eligible installation is 20 years.

** The five years apply to facilities achieving 150 per cent of the reference installation (roughly equals a turbine with an average wind speed of 5.5 m/sec). For other facilities the higher rate is extended for two months for every 0.85 per cent that the facility remains below the 150 per cent of the reference facility. For re-powering projects, the duration of the special rate is longer.

*** Longer if distance from the shore exceeds 12 nautical miles or if water depths are at least 20 metres or more.

This differential cost triggered investments of approximately euro 5 billion in 2005. Total turnover in the field of electricity provision from renewable energies amounted to euro 8.7 billion and according to industry figures there are already around 170,000 people working in the renewables sector in Germany. Now if average cost of the displaced electricity of 4.2 euro cent/kWh is subtracted from the average cost for electricity covered by the EEG for 2005, *i.e.* 9.53 euro cent/kWh, this results in additional cost of 5.33 euro cent/kWh for electricity generation. Multiplied by the total volume of the electricity covered by the EEG, it would amount to the additional cost referred to earlier of around euro 2.4 billion. This corresponds to 0.56 euro cent/kWh shared across total electricity consumption. The monthly additional costs for an average household with an electricity consumption of 3,500 kWh per year currently total around 1.60 euro—in Germany the price of 750 grams of bread. While the costs of conventional electricity generation are clearly on the increase, most of the feed-in fees laid down in EEG are decreasing every year. Against this background, it has been predicted that despite continuously and clearly rising volumes of electricity covered by EEG the monthly costs for a household resulting from the Act will only increase to a maximum of euro 2.80/month by the middle of the next decade and will then decrease.[35]

3.2.7 Eco-tax

With the ultimate objective of reducing energy consumption and the resulting emissions, and to promote the development of environmentally sound products and technologies, on 24 March

1999, the German Bundestag (Parliament) passed the "Law Introducing the Environmental Fiscal Reform", commonly known as Eco-tax, which entails additional excise duties on several energy products. Whereas the tax proceeds are mainly being used to reduce pension insurance contributions, a small portion was reserved for renewable energy subsidies. The revenue from the Eco-tax is used to support the construction of systems to generate heat from renewable energy sources. From an amount of around euro 102 million in 1999 the subsidies for renewable energy sources increased to euro 250 million in 2003.[36] The financing of the renewable energy systems is subsidised by various loan programmes which are handled by the Reconstruction Loan Corporation (Kreditanstalt für Wiederaufbau, KfW).[37] These include the "KfW Environment Programme", the "ERP Environment and Energy-Saving Programme", "Generate Solar Electricity", "Ecological Building", and "Modernisation of Living Spaces". In addition, a portion of the consulting costs are reimbursed for those who use the detailed energy advisory service for older residential buildings.[38]

3.2.8 Bio-Fuels

Share of biofuel in road transport was 3.6 per cent in 2005 which Germany plans to increase to meet EU directives. Since 1992, there has been a mineral oil tax exemption for biodiesel in the country. From 1 August 2006, 10 euro cent tax per litre has been imposed on biodiesel sold at petrol stations and a 15 per cent tax per litre on biodiesel used for blending at refineries. An introduction of quota for biofuels is scheduled from 2007 onwards, with overall additionality quota of 5.7 per cent in 2009. This apart, starting from 1 January 2007, biofuels with quota legislation is subject to mineral oil taxation. Until the end of 2009, tax exemptions for pure biofuels will be maintained on top of quota regulation. The second generation biofuels will be subject to degressive tax exemption up to 2015. It is expected that these measures will create, among others, an enabling condition for advanced biofuel research.

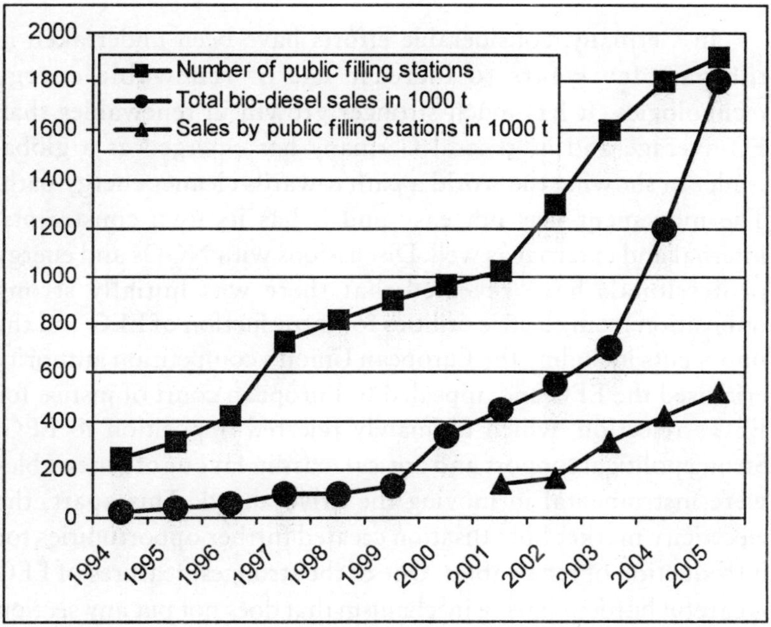

Figure 3.2 Development of the Bio-diesel Market in Germany[39]

3.2.9 General Observations and Conclusion

Germany has relied on a combination of five primary policy instruments for the promotion of renewable energy: direct investment in R&D; direct subsidies; government-sponsored loans; tax allowances; subsidies for operational costs/feed-in tariffs. Over time, Germany's approach regarding renewable energy has shifted from publicly-financed R&D towards policies promoting application and implementation of new technologies in the market place. This transition has occurred gradually and as a result of a number of domestic and international political developments since the 1980s. EU Directives mandating aggressive renewable energy deployment, pressure from domestic environmental advocates and legislators, the phase-out of nuclear power, and commitments to international treaties such as the Kyoto Protocol, all have been important factors in the evolution of renewable energy policy in Germany.

In Germany, considerable efforts have been undertaken in the past few years to increase use of renewable energy technologies. It has much stronger growth in renewables than EU average and in general Germany has emerged as a global leader in showing the world a path towards cleaner energy path. The movement was not easy and it has its own constraints, internal and external as well. Discussions with NGOs and energy professionals have revealed that there was initially strong opposition from electric utilities for introduction of EEG and the opponents including the European Union's competition authority criticised the EEG and appealed to European court of justice for EEG's rejection, which ultimately rejected opposition to EEG. Strong political support and critical mass in favour of renewables were instrumental in moving the drive ahead. This apart, the electricity market liberalisation created further opportunities for introduction of renewables. One of the strongest features of EEG is careful burden sharing mechanism that does not put any section of society at disadvantageous position.

German experience is of immense interest for other countries. The current policies and incentives not only create a good financial framework but also create an environment for continuous technology improvement. The most noteworthy feature is that the policy not only captures the specific requirements of a particular technology, but goes further into various sub categories. Increased decarbonisation of energy sources is one of the most important challenges that industrialised countries have. German example has shown that given a proper policy framework renewables are not only able to penetrate the conventional energy markets but create numerous social and economic opportunities. However, the shift towards renewable energy would not be automatic. The inertia that resists change is too great and the barriers include the lack of knowledge about ultimate benefits from application of renewable technologies, the lack of capital among potential investors and preconditioned notions of policy makers. As such there could not be just one single way of using the potential of increased renewable energy use, rather, a basket of policy measures suiting to different country circumstance would be required. As described above, Germany has witnessed some successful approaches.

3.3 Renewable Energy in the United States

The United States has the largest economy in the world, with a population of 298 million in an area almost two and a half times the size of the European Union.[40] Since 1970, total primary energy supply in the United States increased at an average annual growth of 1.2 per cent with oil the dominant fuel in primary energy supply. The total renewable supply increased from 57 Mtoe in 1970 to around 104 Mtoe in 2003, corresponding to an annual average growth rate of around 1.8 per cent. Whereas biomass and hydropower accounted for about 85 per cent, the share of wind/solar was only around 2 per cent of the total renewable supply. Total primary energy supply (TPES) in the United States was 2280 Mtoe in 2003.[41]

In 2004, total net power generating capacity in the United States was 963 gigawatts and the net generation of electricity 3,971 terawatthours (TWh) of electricity.[42] About 71 per cent of United States electricity was generated from fossil fuels, coal dominates with around half of the total generation, about 20 per cent from nuclear power, another 6.5 per cent from hydroelectric facilities. Among other renewables, which contributed 2.3 per cent in the total electricity mix, the share of biomass was highest with 67 per cent followed by geothermal and wind. Solar thermal and photovoltaics together accounted for less than 1 per cent of U.S. non-hydro renewable generation.[43]

3.3.1 Research and Development

Government investments in energy R&D (including research, development, demonstration and deployment) are intended to accelerate the development and introduction of technologies and practices that provide social benefits, such as increased energy security, reduced energy costs or reduced pollution associated with energy use. The United States spent a total of US $ 106 billion on government energy R&D between 1974 and 2002. In this period, 10.4 per cent of total R&D budget was allocated to renewable energy R&D. US instituted aggressive R&D programmes for renewables in 1974, including the Solar Energy Research Act, which authorized funding for solar and wind research and development, the Geothermal Energy Research, Development and Demonstration Act and Solar Heating and Cooling Demonstration Act. The Wind Energy Systems Act, 1980 provided funding for wind RD&D.[44]

The Department of Energy (DOE) renewable energy programmes are implemented in several ways: through direct funding of R&D at national laboratories, through grants and cooperative agreements with universities, and through various forms of financial and technical assistance to industry partners. In general, the industry partnerships, a prominent part of renewables R&D funding since the mid-1980s, are cost-shared; that is, the industry partner provides a portion of the funding or other resources needed for the work. The cost-sharing can be in the form of direct financial contributions towards the costs of the R&D, or it may be "in kind," meaning that a value is ascribed to some facility or equipment that the industry provides for the effort, or, in the case of the government, the industry partners' use of DOE's national laboratories. The degree of cost-sharing is negotiated and depends on many factors, such as the technology's technical risk (the higher the risk, the greater the government's share) and nearness to commercial readiness, the degree to which the government believes that industry has incentives to do the work on their own, and the size of the industry partner. Small businesses are sometimes provided a higher share of costs than large businesses, at least in early stages of technology development. Although, national laboratories negotiate licensing fees or royalties when they license a technology they have developed to industry, the renewable energy R&D program itself has not required any repayment or royalties from industry partners when jointly-funded R&D leads to commercial success. In partnering programs, DOE also works with the ultimate product consumer to determine desired product characteristics and feeds this information back to its partners.

During recent years, US government has taken many initiatives for developing new technologies. In 2003 US announced the Hydrogen Fuel Initiative, a US $1.2 billion commitment over 5 years to accelerate hydrogen related research to overcome obstacles in taking fuel cell vehicles from the laboratory to the showroom.[45]

In July 2006, US issued a roadmap for new biomass to biofuels research. It aims to accelerate the development of technological breakthroughs needed to efficiently produce cellulosic ethanol. The ultimate goal is to make ethanol a practical and cost-competitive alternative by 2012 (at US $1.07/gal) and displacing 30 per cent (60 billion gallons) of gasoline by 2030. The roadmap identifies biotechnology breakthroughs that have the potential to increase the national biomass supply. The proposed strategy includes: (a) breeding of biomass crops to increase the quantity of biomass per acre, grow better on marginal lands, be more drought and pest-tolerant, and be less costly to harvest; and (b) use of biotechnology which is expected to lead to the breeding of biomass crops with characteristics that make them more easily converted to ethanol, such as the substitution of cellulose. The work involved in this process would include identification of genes involved in the synthesis of cell-wall molecules and higher structures; reactions performed by the multitude of enzymes involved; design principles of cell walls; and factors controlling the amounts, composition, and structure of polymers and polymer matrices.[46]

In September 2006, US announced US $170 million over three years for cost-shared, public-private partnerships to advance solar energy technology, as a part Solar America Initiative (SAI). The SAI aims to bring down the cost of solar energy systems to make them competitive with conventional electricity sources in the U.S. by 2015. The goal of the projects funded by the solicitation is to reduce photovoltaic (PV) costs from 13-22 cents/kWh today, to 9-18 cents/kWh by 2010. It will focus on development, testing, demonstration, validation, and deployment of new PV components, systems and manufacturing equipment. It will be industry-led and may include one or more companies, universities, national laboratories, and/or non-governmental organizations.[47]

In absolute terms there has been impressive development in renewable energy technology development and deployment if compared with other parts of the world. The US have taken a number of initiatives in recent past to develop renewable energy technologies. However, one particular shortcoming is lack of continuous institutional support and linkage between research, development and demonstration investment and future renewable energy production costs. Without institutional support, emerging renewable energy technologies are limited by their financial costs from penetrating the commercial market.

3.3.2 Regulatory Measures and Financial Incentives

In the United States, energy policies are the product of both individual State and Federal policies. In addition to US federal government, States, local authorities and utilities have been providing a number of incentives for renewable energy. They also have implemented various incentives relating to rules, regulations, and policies.[48] Federal policies used to promote renewable energy have included financial incentives, regulatory measures, and research and development (R&D) programs. Among all the States, California has the most active and encouraging renewable energy policies.

In response to energy security concerns of the mid-1970s, the United States passed the National Energy Act of 1978 (NEA), which sought to decrease the Nation's dependence on foreign oil and increase domestic energy conservation and efficiency. A major regulatory mandate that has encouraged renewable energy, the Public Utility Regulatory Policies Act of 1978 (PURPA), was established as a result of the NEA.[49]

PURPA was the most significant section of the National Energy Act in fostering the development of facilities to increase electricity production from geothermal, biomass, waste, solar and wind energy sources. It opened the door to competition in the U.S. electricity supply market by requiring utilities to buy electricity from qualifying facilities (QFs). QFs are defined as non-utility facilities that produce electric power using cogeneration technology, or power plants that use renewable energy sources. The qualifying facilities were also exempted from some of the State and Federal regulations that applied to utility generators.

PURPA requires utilities to purchase power from certain "qualifying" non-utility producers, especially small (below 80 MW) renewables based electricity production at avoided cost rates. Avoided cost is defined as the cost to the utility to generate or otherwise purchase electricity from another source. The Federal Energy Regulatory Commission (FERC) delegated the authority for the determination of avoided cost to the States. This resulted in a number of different regimes, and in several States including California, avoided cost purchase contracts were very favourable to non-utility generators. California based its avoided cost calculations on forecasts of natural gas and oil prices, which were higher than prices actually turned out to be, resulting in favourable investment conditions for renewable power[50]. However, in 1995, the Federal Energy Regulatory Commission (FERC) took responsibility for interpreting "avoided cost," directly linking it with the costs a utility would incur either generating the power directly or purchasing it from another supplier. This interpretation resulted in lower avoided costs than the interpretations of some States, including California.[51]

The United States has also used financial incentives to try to spur the growth of renewable energy. The 1978 Energy Tax Act (ETA) constitutes a programme of tax credits for households and businesses purchasing alternative energy equipment. Residential energy (income) tax credits for solar and wind energy equipment expenditures were set at 30 per cent of first US $ 2000 invested and 20 per cent of next US $ 8000. Another incentive was creation of 10 per cent business energy tax credit for investments in solar, wind, geothermal, and ocean thermal energy installation. In addition, ETA created an excise exemption of US $ 0.04 per gallon of blended gasoline for alcohol fuels. Although the level of these tax credits changed over time until their expiration in 1985, the fundamental policies were developed with the passage of the ETA.[52]

The most important law promoting renewable energy in the 1990s was the Energy Policy Act (EPACT) of 1992. EPACT provided three type of incentives: (a) 10 per cent investment credit for most solar technologies and geothermal; (b) a 10-year 1.5 cents per kWh inflation-adjusted production tax credit (PTC) for tax-paying privately and investor-owned wind projects and closed-loop biomass plants brought online before 1 July 1999; and (c) a renewable energy production incentive of 1.5 cents per KWh for publicly-owned utilities. The incentive expired in 1999, but has since been renewed from time to time and last the legislation extending the PTC was included in a large package of tax incentives in the Energy Policy Act of 2005. It provides a two-year extension (through December 31, 2007) of the 1.5 cent/kWh credit for wind, solar, geothermal, and "closed-loop" bioenergy facilities (Adjusted for inflation, the 1.5 cent/kWh tax credit is currently valued at 1.8 cents/kWh). Other technologies, such as "open-loop" biomass, incremental hydropower, small irrigation systems, landfill gas, and municipal solid waste (MSW), receive a lesser value tax credit. This latest extension increased the number of renewable technologies that are covered by the incentive.

Combined with a growing number of states that have adopted renewable electricity standards, the PTC has been a major driver of wind power development in USA. However, the continuing "on-again/off-again" of PTC has contributed to a boom-bust cycle of development. The cycle begins with the wind industry experiencing strong growth in development around the country during the years leading up to the PTC's expiration. Lapses in the PTC then caused slow down in the implementation of planned wind projects. When the PTC is restored, the wind power industry takes time to regain its footing, and then experiences strong growth until the tax credits expire.[53]

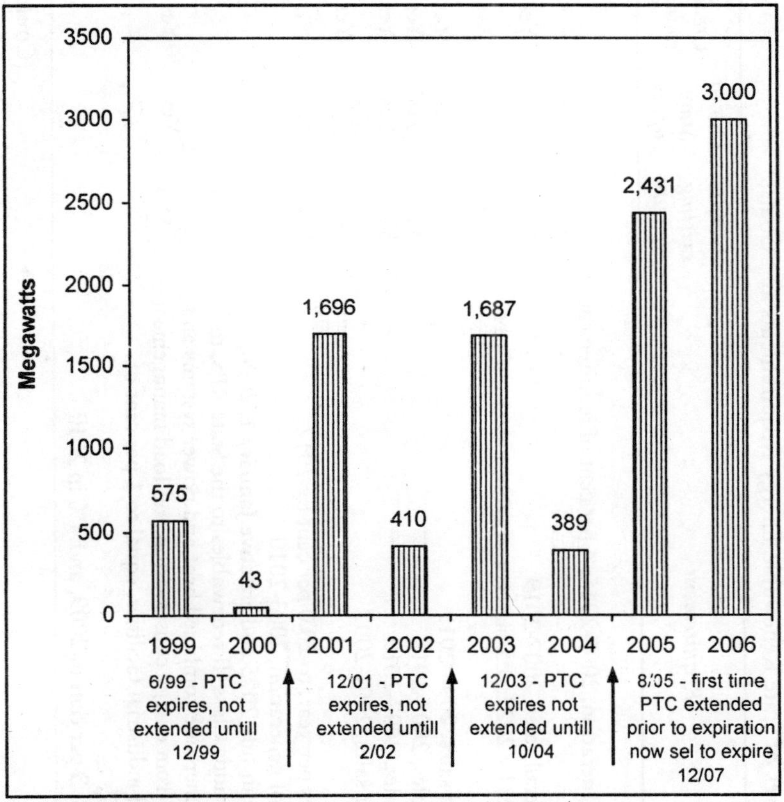

Figure 3.3 US Wind Power Capacity Addition 1999-2006

Section 206 of the Energy Policy Act of 2005 (EPACT 2005) provides for the establishment of a rebate program for expenditures made to install renewable energy systems in connection with a dwelling unit or small business. The amount of the rebate is 25 per cent of the expenditures for qualifying equipment made by the consumer or US $3,000, whichever is less.[54]

Another major policy initiative which have spurred renewable energy development is adoption of renewable electricity standard—a policy that requires electricity suppliers to gradually increase their use of renewable energy. Over 20 US states have adopted a renewable electricity standard.[55]

Table 3.4 Basic Features of State Renewable Energy Requirements and Goals

State	*Year enacted*	*Requirement*	*Accepts existing capacity*	*Out of State supply*	*Credit trading*
Renewable Portfolio Standards					
Colorado	2004	3-10 per cent of generation, 2007-2015; 4 per cent of requirement must be solar	Yes	Yes	Yes
Delaware	2005	1-10 per cent of retail sales, 2007-2019	Yes	Yes	Yes
District of Columbia	2005	11 per cent of sales by 2022; 3.5 per cent of requirement must be solar	Yes	Yes	Yes
Maryland	2004	3.5-7.5 per cent of sales, 2006-2019	Yes	Yes	Yes
Montana	2005	5-15 per cent of sales, 2008-2015	Yes	Yes	Yes
Rhode Island	2004	3-16 per cent of sales, 2007-2019	Yes	Yes	Yes
Arizona	1996	0.2-1.1 per cent of sales, 2001-2007	No	Solar only	Yes
California	2002	+1 per cent of sales per year, to 20.0 per cent by 2017	Yes	Yes	No
Connecticut	2003	6.5-10.0 per cent of generation, 2003-2010 Change was made in July 2005 and effective January 1, 2006, Public Law 05-01 adds Class III renewables to the State RPS, to include new customer-side combined heat and power systems and electricity savings from energy conservation and load management at commercial and industrial facilities, equal to 1 per cent of generation in 2007, 2 per cent in 2008, 3 per cent in 2009, and 4% in 2010	Yes	Yes	Yes

(*Contd.*)

State	Year enacted	Requirement	Accepts existing capacity	Out of State supply	Credit trading
Maine	1997	30.0 per cent of sales by 1999	Yes	Yes	Yes
Massachusetts	1997	1.0-4.0 per cent of sales, 2003-2009	No	Yes	Yes
Nevada	2001	5.0-15.0 per cent of sales, 2003-2013; 5 per cent of requirements must be solar			
		A change was made in June 2005.			
		Assembly Bill 03 increases overall renewables requirement from 5-15 per cent of sales 2003-2013, to 6-20 per cent, but (a) delays compliance by 2 years to 2005-2015, and (b) permits up to one-quarter of the requirement to be met by efficiency measures reducing electricity use.	Yes	Yes	Yes
New Jersey	1999	3.0-6.5 per cent of sales, 2001-2008	Yes	Yes	Yes
New Mexico	2002	5.0-10.0 per cent of sales, 2006-2011	Yes	Yes	Yes
Wisconsin	1999	0.5-2.2 per cent of sales, 2001-2011	Yes	Yes	Yes
Mandates					
Iowa	1983	105 megawatts (no set date)	No	No	No
Minnesota	1994	1,125 megawatts wind by 2010 + 125 megawatts biomass			
		A change was made in May 2005.			
		Statute 216B.243 links compliance with the State's renewable energy goal of 10.0 per cent of electricity sales (by power producers other than Xcel Energy) to obtaining a certificate of need for new transmission or generation capacity.	No	Yes	No

(Contd.)

State	Year enacted	Requirement	Accepts existing capacity	Out of State supply	Credit trading
Texas	1999	400-2,000 megawatts, 2003-2009 A change was made in August 2005. Senate Bill 20 increases overall renewable energy requirement from 2,000 megawatts of new renewable capacity by 2009 to 5,880 megawatts by 2015, including a non-mandatory target of at least 500 megawatts from sources other than wind.	No	Yes	Yes
Wisconsin	1997	50 megawatts by 2000	No	No	No
Goals					
Hawaii	2001	9.0% of sales by 2010 A change was made in June 2004. Senate Bill 2474 changes the goal of the State RPS, from 9 per cent of sales by 2010 to 20 per cent of sales by 2020, and includes ocean technologies, electricity conservation, and some cogeneration.	Yes	NA	No
Illinois	2001	15.0% of sales by 2020A change was made in July, 2005. An Illinois Commerce Commission resolution adopts a sustainable energy plan that replaces the State renewable energy goal of 15 per cent of sales by 2020 with an RPS requiring the State's largest electric utilities to begin supplying 2 per cent renewable energy to Illinois customers by January 1, 2007, increasing by 1 per cent annually to 8 per cent by 2013; at least 75 per cent of the requirement must be from wind power.	No	No	No
Minnesota	2003	1.0-10.0% of sales, 2005-2015	NS	Yes	Yes

(*Contd.*)

State	Year enacted	Requirement	Accepts existing capacity	Out of state supply	Credit trading
Pennsylvania	1998	Individual agreements with five utilities A change was made in November 2004. Senate Bill 1030 changes individual utility goals to RPS requiring 5.7 per cent of sales in 2007, increasing to 18% in 2020 (with solar increasing to at least 0.5 per cent of sales); RPS includes waste coal, coal gasification, and demand-side management and includes both credit trading and some capacity from out-of-State suppliers in interconnected areas.	NS	NS	NS
New York	2004	25 per cent of generation by 2013	Yes	Yes	No
Vermont	2005	All growth, up to 10 per cent of total sales, 2005-2012; goal becomes mandatory if not met by 2012	Yes	-	-

While EPACT significantly improved the economics of windpower, another U.S. policy, implemented thus far at a State level, has been more beneficial to the installation of solar photovoltaic generation. This policy is net metering, which allows small producers of renewable energy from selected sources to sell their power back to the grid.[56] For those consumers who have their own electricity generating units, net metering allows for the flow of electricity both to and from the customer through a single, bi-directional meter. With net metering, during times when the customer's generation exceeds his or her use, electricity from the customer to the utility offsets electricity consumed at another time. In effect, the customer is using the excess generation to offset electricity that would have been purchased at the retail rate. Under most state rules, residential, commercial and industrial customers are eligible for net metering, but some states restrict eligibility to particular customer classes. The buyback rate is determined by law and is frequently equal to the retail electricity rates, or sometimes slightly less than retail rates.[57] Net metering programmes are designed for small electricity customers (residential or small commercial) who produce their own power to bank power on the grid in times of surplus and draw down from the grid in times of need. As of September 2006, net metering was available in 40 States and the District of Columbia.[58] Most States set size limits on systems for net metering eligibility with many States having capacity limits around 25 kilowatts (kW), though limits vary from 10 kW in New Mexico to 2,000 kW in New Jersey.[59]

3.3.3 California

California is the most populous state in the United States, with about 35 million people spread out over 150,000 square miles. California produced 225 billion kWh of electricity in 2005 and imported 22 per cent of its electricity needs.[60] Of the electricity generated within the State, 55 per cent was generated from fossil-fuel sources, 16 per cent from nuclear power plants, 18 per cent from large and small-scale hydropower plants, and about

11 per cent from other renewables. Most of the non-hydro renewable power was generated by geothermal energy, with smaller amounts from biomass, wind energy, and solar.[61]

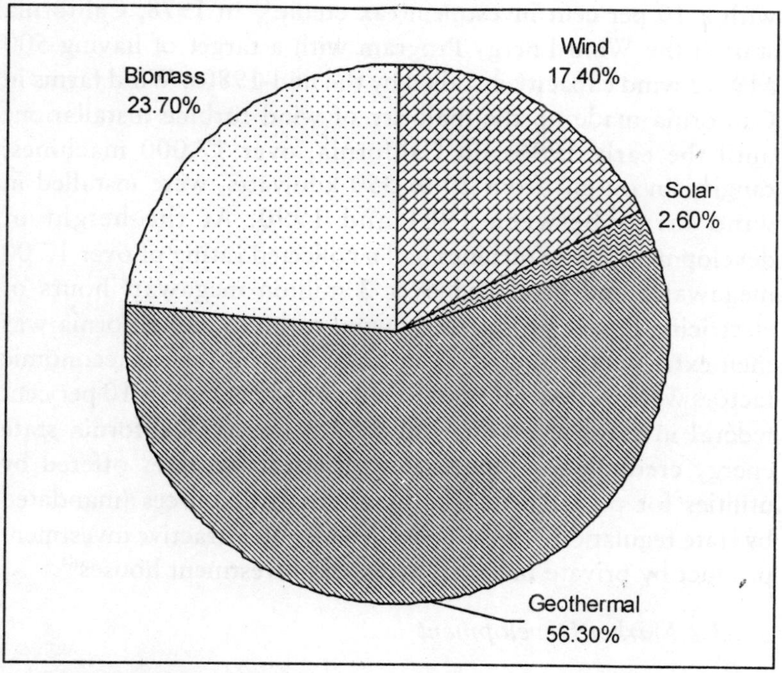

Figure 3.4 Non-hydro Renewable in California 2005 (Total 225 billion Kwh)

California is also a national leader in the development of renewable resources. Over the past 30 years, California has built one of the largest and most diverse renewable generation portfolios in the world. In 2002, California established its Renewable Portfolio Standard programme, with the goal of increasing the percentage of renewable energy in the state's electricity mix to 20 per cent by 2017. The *2003 Energy Report* recommended accelerating that goal to 2010, and the *2004 Energy Report Update* further recommended increasing the target to 33 per cent by 2020. The Energy Action Plan supported this goal.[62]

3.3.3.1 Development of Renewable Energy Policies in California

California began providing tax incentives for installing renewable energy technologies—particularly solar energy devices—in 1976 with a 10 per cent investment tax credit.[63] In 1978, California started the Wind Energy Program with a target of having 500 MW of wind capacity installed by the mid-1980s. Wind farms in California made up the majority of wind turbine installations until the early 1990s. In California, over 17,000 machines, ranging in output from 20 to 350 kilowatts, were installed in wind farms between 1981 and 1990. At the height of development, these turbines had a collected rating of over 1,700 megawatts and produced over 3 million megawatt hours of electricity. Power produced by wind turbines in California was then extremely attractive to utilities. Among the key economic factors were the federal energy credit of 15 per cent, a 10 per cent federal investment credit, and a 50 per cent California state energy credit. These, together with attractive rates offered by utilities for power produced by alternative sources (mandated by state regulations), were packaged into an attractive investment product by private financial firms and investment houses[64].

3.3.3.2 Market Development

In the early 1980s, California moved from funding strict R&D projects for wind energy to focusing on demonstration projects for wind turbines. These demonstration projects resulted in design improvements that helped to bring costs down for wind energy developers. In 1996, California established net metering laws for residential customers who installed solar energy or small-scale wind energy systems of up to 1 MW.[65] The original law applied to wind-energy systems, solar-electric systems and hybrid (wind/solar) systems. In September 2002, biogas-electric facilities up to 1 MW were also included to net meter under a pilot programme until December 31, 2009. Utilities are only required to enroll customers in net metering up to one-half of a per cent of peak electricity demand, a provision which limits the financial burden of net metering on the utilities.[66] California does not allow any new or additional demand charges, standby charges, customer charges, minimum monthly charges, interconnection charges, or other charges that would increase an eligible customer-generator's costs beyond those of other customers in the rate class to which the eligible customer-generator would otherwise be assigned. The California Public Utilities Commission (CPUC) has explicitly ruled that technologies eligible for net metering are exempt from interconnection application fees, as well as from initial and supplemental interconnection review fees.[67]

When California's RPS was enacted on September 12, 2002 (SB 1078), it required retail sellers of electricity to purchase 20 per cent of their electricity from renewable resources by 2017, and was already the most aggressive RPS in the country. Further, the California Energy Commission and CPUC accelerated this goal of 20 per cent renewables to 2010 and set the state's 2020 goal at 33 per cent. All non-hydro renewable sources are eligible, as well as small-scale hydropower and municipal solid waste if it is not combusted. Under the RPS, retail sellers of electricity are required to increase their procurement of eligible renewable-energy resources by at least 2 per cent per year, so that 20 per cent of their retail sales are procured from eligible renewable energy resources by 2010.[68]

In April 2004, through an Executive Order California intended to jump start the use and operation of hydrogen-fueled vehicles in California. It called for a hydrogen highway network, a public/private partnership that will "support and catalyze a rapid transition to a clean, hydrogen transportation economy in California, thereby reducing our dependence on foreign oil, and protecting Californian citizens from health harms related to vehicle emissions." The *Hydrogen Highway Blueprint Plan*, released in May 2005, calls for a dramatic increase in the use of hydrogen-fueled vehicles and a network of hydrogen fueling stations and other infrastructure in three phases. The first phase calls for 50 to 100 fueling stations and 2,000 vehicles by 2010. It also promotes increased renewable resource use with the goal to use 20 per cent renewable resources for both the energy source and feedstock used in hydrogen production by 2010. [69]

The present Renewable Energy Programme of California Energy Commission is organized under three categories; (a) Existing Renewable Resources—supports market competition among in-state existing renewable electricity facilities through varying incentives; (b) New Renewable Resources—encourages prospective renewable electricity generation projects built in California through fixed production incentives; and (c) Emerging Renewable Resources—stimulates renewable technology market growth by providing rebates to purchasers of on-site renewable energy generation while effecting market expansion.[70] Besides the Renewable Energy Programme, public goods charge funds were also being directed towards research in renewable energy technologies. Those funds are allocated under the Public Interest Energy Research (PIER) Programme's Renewable energy Programme Area.

Beginning in 2007, the California Energy Commission manages US $350 million targeted for new residential building construction through the New Solar Homes Partnership. The New Solar Homes Partnership is in response to the Californian Governor's Million Solar Roofs Initiative to create 3,000 megawatts of solar electricity by the end of 2017. Beginning January 1, 2007, the California Energy Commission manages a program of about $400 million, complementing the programme approved by the California Public Utilities Commission (CPUC). The Energy Commission works with home builders and the building industry to accelerate the growth of photovoltaics (PV) in new home subdivisions, whereas the CPUC focues on solar installations on existing residential and commercial buildings.[71]

The California Solar Initiative (CSI), created by the California Public Utilities Commission in January 2006, also begun in January 2007 and provides a total of $2.9 billion over ten years through the Emerging Renewables Programme which provides rebates to consumers who install qualifying renewable energy systems on household's property. The rebate level effective from 1 July 2006 is as under.[72]

Table 3.5 Rebates Available for Emerging Renewable Systems (effective July 1, 2006)

Technology Type	*Size Category*	*Rebate Offered**
Photovoltaic (Solar cells)**	Less than 30 kilowatts	\$2.60 per Watt
Solar Thermal Electric Fuel Cells using a renewable fuel***	Less than 30 kilowatts	\$3.00 per Watt
Wind	First 7.5 kilowatts	\$2.50 per Watt
	Increments between > 7.5 kW and < 30 kW	\$1.50 per Watt

Note: * Rebates for owner installed systems are discounted by 15 per cent.

** Applicants may choose to receive incentive payments based on actual system performance instead of rebates.

*** Fuel cells that operate on non-renewable fuels and are used in combined heat and power applications, may be eligible for rebates at a later date when funds from other sources, such as the Self-Generation Incentive Programme, are no longer available.

3.3.3.3 Conclusion & General Observations

Examination of California's policies towards renewable energy, along with Federal statutes, provides an understanding of historical US renewable energy development. California is the most active US state in renewable energy development and has had a long history of support for renewable energy. Until the early 1990s, nearly all growth in non-hydro renewable capacity in the United States took place in California. In more recent years, other States have begun pursuing policies to increase non-hydro renewable generation, particularly States are developing Renewable Portfolio Standards (RPS).

The non hydro renewable share of US electricity generation has stagnated over the past quarter century at around 2.2 per cent. According to the Energy Information Administration (EIA) of the US Department of Energy, the share of non hydropower renewables will be 4.3 per cent in 2030. Electricity generation from biomass increases from 0.9 per cent of total generation in 2004 to 1.7 per cent in 2030. Generation from wind power increases from 0.4 per cent of total generation in 2004 to 1.1 per cent in 2030. Grid-connected solar generation remains at less than 0.1 per cent of total generation through 2030. These projections accept current federal and state policies already in place—including existing state RPSs and grants, tax credits, and other forms of direct financial assistance—while anticipating higher natural gas prices and relatively lower coal costs.[73]

3.4 Renewable Energy in Japan

Japan is an island eastern Asian nation with a population of around 127 million people and land area of 377,000 square kilometres. It is the second most technologically powerful economy in the world after the US and the third-largest economy in the world after the US and China, measured on a purchasing power parity (PPP) basis.[74] In 2003, the total primary energy supply was 517 Mtoe and the share of non-hydro renewable was 2.1 per cent.[75]

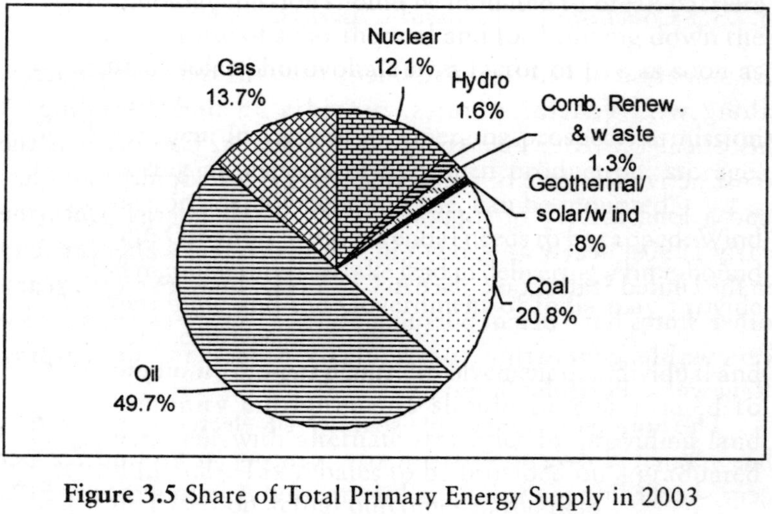

Figure 3.5 Share of Total Primary Energy Supply in 2003

By 2005, Japan's total renewable power installed capacity including small hydro was around 6000 MW.[76] Japan generated 1,047 billion kWh (TWh) of electricity in 2003, of which 66 per cent was generated from fossil fuels, 23 per cent from nuclear power plants, 9 per cent from hydropower and the balance 2 per cent from non-hydro renewables.[77] Major share of other renewables originate from waste and biomass. Solar and wind power account for a very small percentage of renewable energy supply.

The oil price crisis in the 1970s led Japan to work towards securing stable oil supplies, promote the development of nuclear power and renewable energy sources, and encourage energy conservation. The emission reduction commitments under the Kyoto Protocol to the United Nations Convention on Climate Change have also been one of the major driver for renewable energy development in Japan.

3.4.1 Policies

In 1974, the Ministry of International Trade and Industry (MITI), from 2001 the Ministry of Economy Trade and Industry (METI), launched the "Sunshine Project," a long-term comprehensive plan for the research and development of new energy technologies (including solar, geothermal, coal gasification/liquefaction, and hydrogen). The "Moonlight Project" was established in 1978 to boost energy conservation efforts. A research, design and development (RD&D) system focusing on global environment technologies was established in 1989. In 1993, the "New Sunshine Programme" integrated the three above-mentioned projects and aimed at sustainable growth and the resolution of energy and environmental problems. The short-term target of the programme by 2000 was to develop PV technology that could produce electricity at a cost competitive with conventional electricity rates.

In 1980, the Japanese government enacted the "Law Concerning Promotion of Development and Introduction of Oil Alternative Energy" and implemented measures for development and introduction of alternatives to oil including renewable energy. In October 1980, the New Energy Development Organization (NEDO; from 1988, the New Energy and Industrial Technology Development Organization) was established under the Law. In 1981, NEDO began conducting wind research under the auspices of the Sunshine Project. From 1974 to 2001, Japan allocated US$ 3.2 billion to renewable energy research and development. On average over the period, PV received 43 per cent of funding and geothermal benefited from 37 per cent of the total research expenditure. In 2001, Japan allocated US$ 128 million for renewable energy RD&D, of which 62 per cent was for PV, 15 per cent for geothermal and 12 per cent for biomass technologies. The RD&D budget for biomass technologies averaged US$ 40 million through the 1990s.[78] Moving away from the comprehensive approach, renewable energy RD&D is now more technology-specific with separate guidelines and goals for each technology. The research focuses on field tests; promotion and implementation.

The special measures law concerning the use of new energy by electricity retailers, passed in 2002, obliges electricity retailers to supply a certain percentage of renewable energy (Renewable Portfolio Standard). The targets have been set annually from 2004 and will continue until 2010 with the ultimate objective of attaining 12.2 billion kWh (TWh) of electricity generation produced by renewable energy by 2010. Qualifying renewables are wind, solar, biomass, small and medium sized hydropower and geothermal energy. This measure allows power companies to meet their obligations by producing power from new generation sources, purchasing allowable generation from others, or trading with other power companies via a renewable energy certificate trading system.[79]

The Japanese government has planned to meet 3 per cent of total primary energy supply based on new energy sources by 2010. The specific renewable energy targets are 4,820 MW of PV, 4,390,000 kilolitres (kl) of solar thermal production, 3,000 MW of wind power, 4,170 MW of waste power generation, 140,000 kl of waste thermal production, 330 MW of solid biomass generation, 670,000 kl of biomass thermal production, 4,940,000 kl of black liquor and some woody waste.[80]

3.4.2 Photovoltaics

Installed PV capacity in Japan accounted for nearly half of world PV capacity in 2005. Capacity increased from 19 MW in 1992 to 1200 MW in 2005, an average annual growth of more than 40 per cent. The rapid increase in capacity has been supported by Japan's growing domestic PV industry. Japan is the largest manufacturer of PV cells in the world. In 2005, its manufacturing capacity for PV cells was about 823 MW.[81] The prices of solar cells, PV modules and PV systems have been steadily decreasing due to considerable government support for research and development. PV cell technology research commenced under the Moonlight and Sunshine Projects in the 1970s and 1980s.

The growth in PV was driven by four actions on the part of the Japanese government, all taken between 1992 and 1994. First, the government provided net metering guidelines to dictate how distributed PV generation would be hooked up to the national electricity grid. Second, the government set an ambitious goal for the development of PV by 2010. Third, the government introduced subsidies to support the installation of PV technology. Finally, the government established solar roofs programme to encourage residential use of PV and to inform people about the benefits of using PV technology. The Residential PV System Monitor Programme initiated in 1994 was renamed "Residential PV System Dissemination Programme" in 1997 to develop the initial residential PV system market. The total number of PV systems installed under these programmes expanded to the scale of 200,000. This programme aims at reducing the cost burden of purchasers of residential PV systems and creating the initial PV market through subsidizing the installation cost for residential PV systems. The subsidies were available for homeowners installing their own PV systems, for suppliers of ready-built houses, and for public organizations to introduce systems into public buildings. Finally, there was also a subsidy for connecting PV systems to low voltage lines on the power grid. Over the course of the programme, the subsidies decreased as the economics of PV systems became more favourable. Under this programme, Residential PV systems were installed at 115, 765 houses, totalling 421.4 MW, from 1994 to 2002. In 2003, PV systems were set up at 52,863 houses totalling 201.4 MW. In financial year 2004-05, 55,842 applications for the installation of PV systems were received as of January 14, 2005.[82]

In June 2004, New Energy and Industrial Technology Development Organization (NEDO) released a roadmap for technological development of the PV system, "PV Roadmap towards 2030". This provided key guiding principles of further deployment of PV systems and development of the PV industry in Japan. PV 2030 Roadmap laid down technical background to achieve 7 Yen/kWh of PV power generation cost, similar cost level as that for electricity rate for industrial use, by continuous enhancement of technological development and milestones for technical development for the target.[83]

Table 3.6 PV Road Map: Technological Targets Towards 2030

Item	*Present Status*	*Target by 2010-2030*
Production cost of PV module	Production: 250 yen/W (2003) Expected development: 14 yen/W (2007)	100 yen/W (2010) 75 yen/W (2020) <50 yen/W (2030)
Durability of PV module	20 years	Service life 30 years (2020)
Silicon feedstock consumption	10-13 g/W	1g/W (2030)
Inverter (power conditioner unit)	30,000 yen/KW	15,000 yen/KW (2020)

Table 3.7 PV Road Map: PV Module Conversion Efficiency Targets *(cell efficiency targets)*

Solar Cell Type	*Present Status*	*Conversion efficiency target (%)*		
		2010	*2020*	*2030*
Crystalline silicon solar cell	13-14.8 (18.4)	16 (20)	19 (25)	22 (25)
Thin-film silicon solar cell	10 (14.7)	12 (15)	14 (18)	18 (20)
"CuInSe" solar cell	10-12 (18.9)	13 (19)	18 (25)	22 (25)
"III-V" solar cell	Concentrator (38.9)	28 (40)	35 (45)	40 (50)
Dye-sensitized solar cell	(10.5)	6 (10)	10 (15)	15 (18)

3.4.3 Wind Power

Wind power capacity increased from 6 MW in 1999 to 1078 MW in 2005.[84] Wind power still plays a minor role in renewables electricity generation. RD&D support for wind power commenced in the early 1990s with wind turbine system control technology development and a wind quality survey (*i.e.*, nationwide wind condition atlas). The development of large-scale wind systems began to receive support in 1991. Since 1995, NEDO has financed wind power field tests, and by 2001 wind power plants were installed in thirty-one areas. The 1997 Law Concerning Promotion of the Use of New Energy defined wind power as a new energy source, making it eligible for incentives offered under the law. Since 1997, NEDO has provided subsidies of one-third of the cost of installing turbines to private sector firms, and of one-half of the cost to local governments. As a result of these subsidies, growth in the wind market has accelerated.

Other sources like geothermal and biomass power are also being promoted in Japan under the overall policy framework for renewable energy market expansion.

3.4.4 New Energy Strategy

In May 2006, METI declared New Energy Strategy.[85] Its one of the major objectives is to establish foundation for sustainable development through a comprehensive approach for energy and environmental issues. To this effect under new energy innovation plan, the targets set include bring down the cost of solar energy generation to the same level as thermal power, promoting the efforts of local-production for local-consumption through biomass energy and wind-powered electricity to improve the self-sufficiency ratio of the energy supply, and turning many of the new vehicles sold in the market into hybrid vehicles along with promoting the introduction of electric vehicles and fuel cell vehicles. The strategy provides the basis for proactive introduction of the new energy related facilities in public establishments, by applying the RPS law, and providing necessary market expansion support such as the application of assistance and taxation systems. It sets a target of forming a substantial industrial structure for

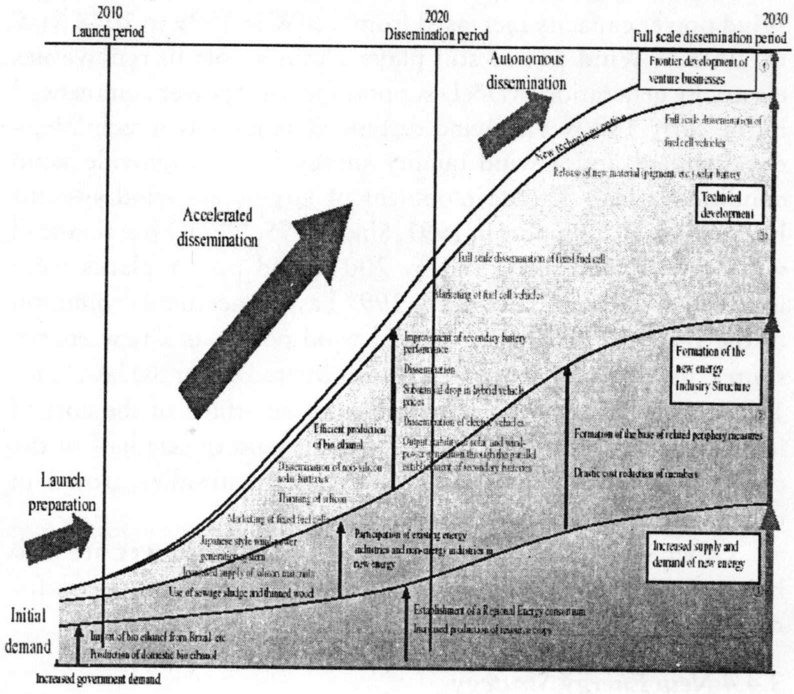

Figure 3.6 Japan's New Energy Strategy: Expansion of the Introduction of New Energy

the new energy industry by developing a group of solar energy generation industries, developing a group of fuel cell and secondary cell industries, and promoting regional businesses based on the local-production and local-consumption of wind-power and biomass.

3.4.5 General Observations

Japan is poor in energy resources and this was the driving force for developing a new energy strategy which not only takes into considerations the geopolitical issues but provides a strong trust for development of renewable energy sources through market development and time bound research and development planning. Its renewable energy strategy includes down the price, drop subsidies and continue growth. Being a high energy utilizing country its focus on renewable is from the climate change considerations with its emission reduction target. The present thrust to renewable energy in coupled with strategic policy intervention to facilitate and accelerate the transition. Efforts towards the introduction of a comprehensive market of renewable energy technologies include full-cost energy pricing, environmental regulations, tax incentives, codes and standards, public education.

3.5 Renewable Energy in China

China is an eastern Asian nation with a population of around 1,296 million people. Its economy has maintained an approximate annual growth rate of 9 per cent during the past 20 years, along with an increasing demand for energy. In 2004, the total primary energy supply was 1609 Mtoe and the share of combustable renewable and waste was 13.7 per cent.[86] China's renewable energy industry has grown steadily over the last decade, and a principal target of technological advancement has been rural and household-scale power systems. Almost 97 per cent of Chinese households have access to electricity.[87] This is one of the reasons why China needs to develop alternative and renewable energy as an additional source of energy supply.

In 2004, China had total installed electricity generating capacity of 391.4 gigawatts (GW), 74 per cent of which came from conventional thermal sources. In 2004, China generated 2,199 billion kilowatthours (Bkwh) and consumed 1,721 Bkwh of electricity. Non-hydro renewable share in electricity generation was from biomass, which was around 2.5 Bkwh.[88] By 2005, the contribution of non-hydro renewable sources was around 3300 MW with 1300 MW from wind and the balance from biomass. According to the State Renewable Energy Medium-and Long Term Development Programme, renewable energy is expected to account for 16 per cent of China's total energy supply by 2020, up from 7 per cent in 2005.

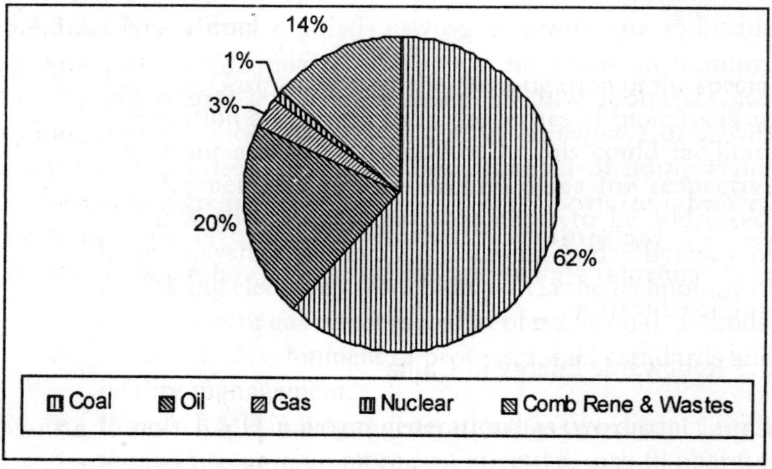

Figure 3.7 Share of Total Primary Energy Supply in 2004

3.5.1 Rural Renewable Energy Programme

By 2006, there remained only an estimated 4-5 million unelectrified households in rural areas of China, representing about 15 million people. There are an estimated 600,000 to 650,000 solar home systems installed in rural areas. China has had major projects for rural electrification in recent years including the Township Electrification programme for electrification of around 300,000 households during 2002-2004 and the Brightness Programme during 2001-2004 that set up 30,000 SHS and 40 village-scale PV systems.[89]

China's Brightness Project was initiated in 1998 and around US$ 50 million were allocated by the China's State Council for the period 2001-2005. The project aimed at providing electricity using both household and village systems and the areas covered include Gansu, Qinghai, Inner Mongolia, Tibet, and Xinjiang provinces. The ultimate objective of the project is to provide electricity for 23 million people in remote areas by 2010 using renewable energy technologies.[90]

Household-scale biogas for rural home lighting and cooking is a wide-spread application in China. About 17 million households have household-scale biogas digesters. Total biogas production was 8 billion m^3 in 2005, including about 1500 industrial scale biogas installations. A typical household-scale digester, sized 6-8 m^3, produces 300 m^3 of biogas per year. A new government programme, started in 2002, provides 1 billion RMB annually as subsidies to farmers who build their own digester. The subsidy is 800 RMB per digester. Some estimate that more than 1 million biogas digesters are being produced each year.[91]

In 2001, China's National Development and Reform Commission [NDRC] launched a renewable energy-based rural electrification program known as Song Dian Dao Xiang, literally "Sending Electricity to Townships." By 2003, 100 townships were covered under the programme and the systems installed included 20 MW from PV, 840 kW from wind, and 200 MW from small hydropower and the government provided US $ 240 million as subsidy. Under the second phase for 2005-2010 electrification of 20,000 villages in China's off-grid western region has been planned.[92] There are about 200,000 stand-alone small wind turbines serving rural households, primarily in Inner Mongolia, with total capacity of about 35 MW (an average of 170 watts each).

3.5.2 Research and Development

The central government subsidies R&D on key renewable energy technologies through National Development and Reform

Commission (NDRC) and Ministry of Science and Technology (MOST) as well as local government. R&D funds on renewable energy offered by MOST during the Ninth Five-Year Plan (1995-2000) period was 60.0 million RMB Yuan. During the Tenth-Five-Year Plan period, it increased to 200 million RMB Yuan. In addition, by MOST, the central government also supports the R&D of renewable energy technologies through national programs including 863 programme (which is for the near-commercialization technologies) and 973 program (which is for fundamental technologies). Therefore, the total fund for R&D on renewable technologies through MOST was 1 billion RMB Yuan. Furthermore, there are some subsidies for demonstration projects and training courses by NDRC, Ministry of Finance (MOF) and MOA.[93]

3.5.3 Renewable Energy Policy

The initial focus of the Chinese renewable energy policy was on the development of rural energy and rural electrification. In 1994, the then Ministry of Power issued guidelines for the construction and management of wind farms. In 2002, in order to increase the attractiveness of wind energy projects in China, the VAT for wind generation equipment was decreased from 17 per cent to 8.5 per cent and income tax for wind projects from 33 per cent to 15 per cent. In respect of small hydro and biogas also reduced VAT of 6 per cent and 13 per cent respectively are applicable.

In an effort to develop large-scale wind farms, effectively bringing down the cost of wind through competition, the Chinese government introduced a wind concessions approach in late 2001. The basic concept of the Wind Power Concession was that the local government will invite international and domestic investors to develop 100 MW wind farms on a potential wind site, through a tendering procedure aimed at bringing down the cost of wind-power generation. In October 2003, two companies were selected through competitive bidding to develop the first large-scale wind concession projects in China. The wind concession lasts for 25 years and the bid price is guaranteed as a feed-in tariff for the first 30,000 full load hours achieved (for a 100 MW project, this amounts to approximately 3 billion kWh). Depending on the site's wind resource, this could cover about 10-15 years. After 30,000 full load hours, the project owner will receive the average local feed-in-tariff in the power market at that time. These two projects required 50 per cent domestic content in turbines. In 2004, the Chinese Government offered three more concession projects of 100-200MW in size, and these concession projects requires 70 per cent domestic content. The NDRC expects to award a total of 20 such projects by 2010.[94]

3.5.4 Renewable Energy Law

The Renewable Energy Law of the Peoples' Republic of China came into effect on 1 January 2006.[95] The authorized text of the Chinese Renewable Energy Law may be seen at http://china.lbl.gov/publications/re-law-english.pdf. The entry into force of the Renewable Energy Law was an important symbolic event and during that year, several renewable energy industries grew rapidly.

The Law established a system for: (a) establishing national targets; (b) grid connection priorities; (c) classifying tariffs for renewable power; (d) sharing cost at national level; (e) renewable energy special fund; and (f) policy on favourable credit and favourable tax treatment. In line with the provisions of the Law the major tasks for regulations' formulation include: (a) national general targets; (b) national renewable energy plan; (c) grid-connection pricing mechanism; (d) cost-sharing measures; (e) economic incentives; (f) financing backup measures in rural regions; (g) renewable energy development special fund; (h) technical criterions for solar integrating building, resource investigation and assessment, grid connection and other national standards.

3.5.4.1 Principles of the Renewable Energy Law

Four key principles were established as the basis for China's renewable energy legislation.

3.5.4.1.1 Combining Government Responsibility and Citizens' Obligations

The Chinese Government has clarified its responsibility to develop renewable energy in the Renewable Energy Law. At the same time, the Government has stated that both companies and individuals will be encouraged, and in some cases obliged, to use renewable energy. For example, the cost sharing system requires all power consumers to pay a surcharge to support the development of renewable energy, and regulations provide for the compulsory use of solar power in some buildings. The principle of combined responsibility lies at the core of the Renewable Energy Law and serves as the basis for all the regulations.

3.5.4.1.2 Combining Government Promotion and Market Guidance

The Chinese Government recognizes its role in building a market, establishing the market rules, and regulating the market. Once a market is established, market mechanisms will efficiently guide and encourage participants to explore renewable energy resources. The Renewable Energy Law contains detailed regulations on responsibilities for renewable energy resources surveys, planning, project construction, industrial development, and economic incentives. Regulations detailing the market rules and how market competition is to be maintained will also be established. These regulations are designed to assist in building the market, as well as encouraging the active involvement of market players in renewable energy development.

3.5.4.1.3 Combining Actual Demand and Future Development

The government's aims for the Renewable Energy Law include, on the one hand, to satisfy existing energy demands, while on the other hand, balancing planning for future energy supply and demand. The Renewable Energy Law is intended to ensure that mature renewable energy technologies, such as, solar, biomass, biogas and hydropower technologies, will be promoted in an effort to solve rural energy access problems. It is also intended to ensure that research and development of future technologies will be strengthened. Economic incentives are intended to create market demand in new technologies.

3.5.4.1.4 Combining Domestic Practice with International Experience

The key provisions of the Renewable Energy Law have combined domestic practice with international experience. For instance, the overall target for renewable energy takes into account the domestic context, including the availability of particular resources, existing manufacturing capacity and the requirements of rural and urban areas. On the other hand, the pricing mechanism and cost sharing system have benefited from the lessons of international experience.

3.5.4.1.5 Division of Responsibilities

Under the institutional framework of China, the central government is responsible for the formulation of the national regulations, which guide individual provinces during the implementation process. Instructions regarding pricing, cost sharing, taxation and the project approvals process are stipulated by the central government for the provincial government to follow.

The Renewable Energy Law requires the relevant central authority to be responsible for drafting these regulations. For example, the National Development and Reform Commission (NDRC) is responsible for energy pricing and planning issues, the Ministry of Finance (MoF) is responsible for economic incentive programs, and the Ministry of Agriculture is jointly responsible with the NDRC for rural biomass utilisation and rural energy development. Technical codes relating to renewable energy in buildings have been issued by the Ministry of Construction (MoC).

3.5.4.1.6 Regulations Issued So Far

Under the Renewable Energy Law, implementing regulations are to be formulated that cover key issues including pricing, grid-connection and incentive policies. So far, seven such documents have been published, as under:

Table 3.8 Regulations under Renewable Energy Law

Provisional Administrative Measures on Pricing and Cost Sharing for Renewable Energy Power Generation (NDRC Price [2006] No. 7)	Sets out the principles for renewable energy power pricing and cost sharing. In particular, it identifies the level of wind and biomass power pricing and clarifies all costs related to renewable energy power that will be covered by the renewable energy surcharge.
Level Regulation (NDRC Price [2006] No. 28-33)	Establishes the tax-exempt renewable energy surcharge (¥0.001 per kWh) payable by end users of electricity. This cost sharing arrangement mandates that end users pay a proportion of the higher cost of providing renewable energy, as well as the cost of connecting renewable energy facilities to the grid.
Provisional Regulation on Renewable Energy Surcharge Balancing (NDRC Price [2007] No. 44)	Identifies the procedure for provincial power utilities to collect the renewable energy surcharge, the methodology for allocating this revenue amongst the provinces, and the role of the monitoring body in this process.
Regulation on the Administration of Power Generation from Renewable Energy (NDRC Energy [2006] No. 13)	Sets out approval procedures for renewable energy projects and further identifies the responsibilities of utilities and power generators. Provides that utilities are obliged to allow renewable energy facilities to connect to the grid.
Guiding Catalogue for Development of the Renewable Energy Industry (NDRC Energy [2005] No. 2517)	Identifies the renewable energy technologies that will be supported by the government and identifies the economic policy instruments that will apply to these.

(*Contd.*)

Provisional Administrative Measures on the Renewable Energy Development Fund (MoF Economic and Construction [2006] No. 237)	Sets out the criteria for the use of the Renewable Energy Development Fund, identifies "priority areas", and provides application and approval procedures.
Regulation Governing the use of the Renewable Energy Development Fund to Promote Renewable Energy Integration in Buildings (MoF Construction [2006] No. 460)	Together with the 'Notice on the approach to appraisement of pilot projects for renewable energy integration in buildings', sets out how the Renewable Energy Development Fund will be used to promote the integration of renewable energy in buildings, the application and approval procedures and the criteria for project selection.
Regulation on the Manage-ment of Bio-Ethanol Projects (MoF Construction [2006] No. 460)	Sets out the policy for bio-ethanol development, imposes stricter market-entrance standards, project management and supervision requirements, and streamlines the administration system.
Regulation of The Construction And Management of Wind Farms (NDRC Energy [2006] No. 1204)	Obliges local government authorities to develop local wind energy development plans (for facilities smaller than (50MW) according to wind resource availability. The wind tariff is still determined by the State Council through a tender process.

3.5.4.1.7 Systems Established by Renewable Energy Law

The Renewable Energy Law has provided the framework for some important legislative initiatives, designed to secure the strategic position and future development of renewable energy. These include:

(a) Renewable energy targets, including both economy-wide and technology-specific targets;

(b) Compulsory grid connection for renewable energy facilities to the State electricity grid;

(c) Power pricing arrangements, including feed-in tariffs and competitive tendering systems, to allow renewable energy to compete with traditional, fossil fuel-powered generation; and

(d) Cost sharing arrangements to divide the costs of renewable energy generation and grid connection equitably amongst utilities and electricity end users.

3.5.4.2 Renewable Energy targets

The overall and technology-specific targets for 2010 and 2020 are as in Table 3.9.

Table 3.9 Renewable Energy Targets

	2010	*2020*
Share of renewable energy in national energy mix	10%	16%
Hydro	180 GW	300 GW
Biomass	5.5 GW	30 GW
Wind	5 GW	30 GW
Solar	500 MW	1.8 GW
Solar collectors	150 million m^2	300 million m^2
Biogas	15 billion m^3	30 billion m^3
Liquid fuel	2 million tonnes	10 million tones

Achievement of the aforementioned targets would result in renewable energy constituting approximately 16 per cent of China's projected energy consumption in 2020.

3.5.4.3 Grid Connection Under the Law

The Renewable Energy Law provides for the compulsory connection of renewable energy generators to the grid, and a regulation has been enacted to give effect to this (NDRC Energy [2006] No.13). The regulation deals with general rules governing grid connection, project management requirements, utility company responsibilities and generator responsibilities. The regulation covers hydropower, wind power, biomass (including forest and agricultural residue, direct combustion and gasification, land fill gas power generation, biogas generation), solar power generation, geothermal power generation, and ocean energy. It emphasises that penalties will be imposed for violations.

There are two elements to the compulsory grid connection system in the Renewable Energy Law:

(a) All energy generated from renewable sources must be purchased; and
(b) Utilities must provide grid-connection services (including constructing grid connections) and related technical support.

The Renewable Energy Law has also provided clear regulations for gas, heat and liquid fuel sourced from renewable energy. These forms of energy generation must be allowed to be connected to the grid if they satisfy the standards of grid connection.

National biofuels standards are currently under preparation. Oil companies must also incorporate biofuels that accord with these standards into their distribution systems. As special monopoly industries, these oil companies have a specific responsibility to renewable energy generators to facilitate their integration into the distribution systems.

3.5.4.4 Price Setting for Renewable Energy

Under the Renewable Energy Law, renewable energy prices (or the mode by which they are established) are to be determined by the price authorities of the State Council in accordance with the principle set out in Chapter 5, Article 19 " being beneficial to the development and utilisation of renewable energy and being economic and reasonable, where timely adjustment will be made on the basis of the development of technology."

3.5.4.5 Methods to Price Renewable Energy

There are typically two approaches for pricing of renewable energy: (a) prices may be determined by the market (supplemented by regulation, such as the implementation of renewables portfolio standards or competitive tendering) or directly by regulation (through the use of feed-in tariffs). Both approaches are used in

China. The Chinese Government has chosen to implement separate pricing laws for each type of renewable energy, using two primary methods: feed-in tariffs "government-fixed pricing" and competitive tendering "government-guided pricing". These laws and regulations aim to offer renewable energy generators a guaranteed power price, coupled with a purchase obligation on utilities, to stimulate the development of the market. So far, only wind, solar and biomass prices have been developed, although a framework methodology for wave, hydro and geothermal power prices is set out in the Renewable Energy Law.

Table 3.10 Pricing Mechanisms for Renewable Energy in China

Type of Energy	*Price-Setting Method*	*Details*
Biomass	Feed-in tariff	Price for coal plus 0.25 Yuan/kWh, decreasing by 2% per year from 2010.
Solar PV and Solar Thermal	Feed-in tariff with government approval of projects	Government first needs to approve each solar power project. If a project is approved, the government will set an appropriate feed-in tariff, on a project-by-project basis. The tariff will be set based on the concept of a "reasonable price".
Wind	Tendering	Government will select potential investors through a competitive bidding process, with power price and domestic content the key criteria.

3.5.4.6 Wind Power-competitive Tendering

Under the competitive tendering arrangements in place for the wind power sector, renewable energy developers must submit proposals for large-scale renewable energy projects of 100MW and above. The government will select potential investors through a competitive bidding process, taking power price and domestic content as the key criteria.

The government will guarantee the purchase of all electricity from the project through long-term power purchase agreements.

The price will be than set during bidding for the first 30,000 MWh, and the average market power price for the remainder of the period (25 years total).

3.5.4.7 Cost sharing—who pays the extra costs?

The Renewable Energy Law frames it in the following terms:

> "The difference between the price of renewable energy power (as determined in Article 19) and the average price of conventional power will be shared through the sale price. Price authorities of the State Council will prepare specific methods for cost sharing.
> Grid connection expenses incurred by grid enterprises in order to purchase renewable power and other reasonable expenses may be included in the power transmission cost and retrieved through the sale price."

Under China's cost sharing arrangements, end users of electricity will be required to pay a tax-exempt surcharge to cover part of the difference between the price of grid-connected renewable energy and local desulphurised coal power, as well as the cost of grid extensions to connect renewable energy power facilities to the grid, and operation and maintenance of off-grid renewable power facilities.

The surcharge is set by the pricing department of the State Council, and calculated according to a formula in the pricing guidelines. The current surcharge is 0.001 Yuan/kWh (on top of the 0.4-0.6 Yuan/kWh charged for residential electricity), collected by grid corporations.

The other component of the excess cost is shared by all energy utility companies nationwide and cannot be passed on to consumers.

3.5.4.8 Investment Incentives for Renewable Energy

The regulations under the Renewable Energy Law will establish incentives to invest in renewable energy in China. The incentive schemes currently being drafted include preferential tax treatment, low-interest loans, and preferential tariffs. The exact forms that these schemes will take and the level of incentives which will be provided are not yet known, although the Renewable Energy Law states that they will apply to renewable energy development and utilisation projects.

3.5.4.9 Renewable Energy Development Fund

To facilitate the implementation of the Renewable Energy Law, the Ministry of Finance issued in June 2006 a regulation concerning a special fund for renewable energy development. Companies (including domestic and foreign-invested enterprises) and individuals may apply to this fund for free financial aid or loan interest discounts. Currently, such funds are available mainly for the development of bio fuels, the promotion and application of solar energy and geothermal energy in construction and building, and the promotion and application of power generation from wind power, solar energy and marine energy.

3.5.4.10 Local Content Requirements

Wind power projects in China must have over 70 per cent of their wind turbine component locally made, and the wind turbine generator (WTG) must be assembled in China. The aim of this requirement is to encourage China's technology and manufacturing industry for wind turbines.

3.5.4.11 Official Review of the Renewable Energy Law

The NDRC carried out the first official government review of the Renewable Energy Law in early 2007. The results of the review, including recommendations from Chinese industry stakeholders, were published on 20 April 2007. The recommendations arising from the review are strengthening of the implementation of the Law and the following the policy Measures:

(1) Improve the pricing mechanisms for wind projects. The existing mechanism provides that the pricing authority of the State Council is to set prices according to tenders submitted, and a bidding process is undertaken by the Government for every project (at a significant cost to both the Government and investors). The review recommended that tender process be conducted only for large-scale wind projects greater than 100 MW, and that small and medium-scale projects adopt the prices set for large-scale projects operating in the same region or nearby.

(2) Implement a renewable portfolio standard as soon as possible to increase market confidence about prices.

(3) Accelerate the formulation of preferential tax and other fiscal policy. In particular, it was proposed that VAT on wind energy facilities be reduced to within a reasonable range to facilitate a reduction in the wind tariffs and maintain local government enthusiasm for wind power development. Implement measures to support research and development into the development of large wind turbines and provide subsidies on domestic turbine procurement and preferential loans (from state-controlled commercial banks). For the solar hot water industry, recommendations include eligibility for preferential tax policies currently available for high-tech, renewable energy and home appliance industries.

(4) Modify the current duty free arrangements for whole wind turbine imports, so that duty free status is only available to those turbine components that cannot be produced domestically, in order to accelerate the development of the Chinese wind turbine manufacturing industry. Furthermore, it is proposed to more strictly apply the local content requirement for developers tendering for large-scale wind projects, and to adopt preferential treatment for the wind power equipment manufacturing sector.

(5) Accelerate the formulation and publication of national renewable energy targets and the long-term plan for

renewable energy, and formulate and publish specific plans for wind energy, biomass, solar energy and other technologies. The purpose of the plan is to provide a 'road map' for development of all sectors in the renewable energy industry. Align policy mechanisms with long-term targets.

(6) Clarify responsibilities for implementation of renewable energy strategies and policies, both nationally and within each province.

(7) Bring forward development timelines for large-scale wind projects where possible to increase the predictability of the wind energy market.

(8) Clarify responsibilities for resource assessment and initiate data collection projects as soon as possible (with wind and biomass as a priority). Assessment of wind resources is to be funded by the NDRC, the National Meteorological Bureau and other departments; biomass resource assessment by the NDRC, the Ministry of Agriculture, the State Forestry Bureau and other departments. The results of the projects should be published as soon as possible to reduce market uncertainty.

(9) Increase funding for research and development into high-tech and industrial equipment technology projects in the renewable energy sector.

(10) Establish a disclosure system to enhance public access to project planning information (including approval status, pricing data etc.) and information on progress in the implementation of the Renewable Energy Law.

(11) Create a centralised energy management department within the national government, with responsibility for energy industry strategy and policy development.

(12) Reform the administrative approvals system to simplify approval procedures for non-hydro projects and to maximise public safety, resource efficiency and environmental protection.

(13) Create special funds for agencies to support the development of intellectual property rights.

(14) Promote the development of industry associations and other non-governmental organisations and encourage their participation in renewable energy policymaking.
(15) Implement projects to raise public awareness of renewable energy, including developing renewable energy demonstration projects in public buildings.
(16) Create voluntary green energy purchase schemes and corporate 'green logos' and environmental ratings.

3.5.4.12 National Climate Change Programme

In June 2007, China published a new National Climate Change Programme. The programme contains several statements relating to the Renewable Energy Law. It states that the measures set out in the programme will assist in "vigorously developing renewable energy" (section 3.3.1). Section 4.1.1(1) of the programme states that China will promulgate the Renewable Energy Law as early as possible, and implement it in a comprehensive manner and further intensify preferential policies to develop and utilize clean and low carbon energy develop supportive regulations and policies, prepare national and local programs for renewable energy development, identify development objectives and integrate renewable energy development into assessment indicator systems for the construction of resource-conservative and environmentally friendly society through legislation and other approaches, [guide and encourage] domestic and international economic entities ... to participate in renewable energy development and utilization.

Furthermore, section 4.1.1(2) asserts that:

> A stable mechanism for [renewable energy] investment will be established through government investment, government concession and other measures. A sustainable and stably expanding market for renewable energy will be fostered, market environment for renewable energy will be improved and obligation of national electricity grids and petroleum sales enterprises under the Renewable Energy Law to purchase renewable energy products will be implemented.

3.5.5 General Observations

China has three existing laws pertaining to energy issues: the Electricity Law, the Energy Conservation Law, and the Renewable Energy Law. The Renewable Energy Law was endorsed unanimously by the National People's Congress. It is one of the most comprehensive law that not only addresses renewable power development issues but takes into consideration other renewables. Provisions for managing a special state fund for renewable energies and measures for giving renewables projects discount loans and preferential tax treatment are indications of the seriousness of the Chinese Government to harness renewables for meeting country's growing energy needs in a sustainable manner.

3.6 An Overall Policy Comparison

Renewable energy policies in EU, Germany, USA, Japan have by and large been coordinated and consistent. China has also a long history of renewable energy programme and policies, but a focused attempt have been made with adoption of renewable Energy Law in 2006. In the United States, policies are mostly enacted at the State level and as such there could not be synchronization with the federal policies. EU and Germany have shown extraordinary level of political commitments to renewable energy that was both consistent and well funded. The varying level of renewable application in these countries is attributed to resource endowments, political and economic systems and electricity prices and most importantly political commitments. Japan and Germany enacted policies expressly for the purpose of promoting renewable energy much before US which initiated the process with EPACT in 1992. Although renewable energy policies in each of the identified countries have some elements of commonality in them, specific emphasis include, Feed-in Tariffs in Germany, Renewable Portfolio Standards (RPS) and Green certificates in Japan, Production and Investment Tax Credits in the US and a policy framework under Renewable Energy Law in China. The policy instruments adopted by EU, Germany, USA, Japan and China are briefly summarized in the Table 3.11.

Table 3.11 A Comparison of Renewable Energy Policies of Reviewed Countries

Country	*Research and Development*	*Renewable Electricity*	*Net metering*	*Financial Incentive/Subsidy*	*Remarks*
EU	Since 1984, research and innovation activities grouped in one programme called the Framework Programme (FP).	Directive 2001/77/EC to increase renewable energy's share to 12 per cent of primary energy consumption and to boost renewable sources to 22 per cent of electric power production by 2010	—	Support to demonstration projects in various EU member countries	EU has set indicative targets by launching 'Campaign Sustainable Energy 2005-2008' in 2005.
Germany	(1974-Present) Early R&D conducted by firms in the aerospace industry and was focused on large turbines (larger than 1MW), which were perceived as having the best chance of becoming marketable	Electricity Feed-In Law (1991) & Renewable Energy Law (2000) further revised in 2004	Net metering is implicit in fee-in-tariff system	Market development programmes supported with low interest loans	German renewable energy producers received higher rates than United States producers

(*Contd.*)

Country	Research and Development	Renewable Electricity	Net metering	Financial Incentive/Subsidy	Remarks
USA	R&D programme, initiated in 1974 initially contracted with companies from the aerospace industry and focused on large-scale turbine technology with technology development programmes for other renewables	Public Utility Regulatory Policies Act (PURPA) required utilities to buy power from renewables producers and purchase price for power was at utilities "avoided cost," which was defined by the states	Net metering policies enacted in 32 States and the District of Columbia	Million Solar Roofs Initiative (1997) which seeks to install a million solar PV or solar thermals systems by 2010	No financing offered at Federal level and RPS Laws enacted in various states. For comparison Germany linked rates to retail prices while California rates were linked to (high) projections of "avoided cost"
Japan	A research, design and development (RD&D) system was established in 1989 and in 1993, the "New Sunshine Programme" launched with the objective of developing PV technology that could produce electricity at a cost competitive with conventional electricity rates. In 2006 a 'New Energy Strategy' with a focus on R&D was launched.	Encourage the use of PV systems in residences. It has set a target of 1.35 per cent of total electricity to be generated by renewables by 2010	Sets parameters for connecting small, distributed PV and wind systems to the grid	70,000 Solar Roofs with investment subsidies (1994) RPS (2002)	

(Contd.)

Country	*Research and Development*	*Renewable Electricity*	*Net metering*	*Financial Incentive/Subsidy*	*Remarks*
China	R&D funds on renewable energy have been provided under five-year plan. R&D and the industrial development of renewable energy is listed as the preferential area for hi-tech industrial development in the national programme.	Whereas feed—in tariff for biomass and PV, wind electricity price will be determined through tendering process.	—	Renewable Energy Law provides for Government support to renewable energy through appropriate financial instruments, including financial support.	The aim is to contribute around 16 per cent of energy from renewables by 2020. For the purpose the Law provides for a stable mechanism for renewable energy investment, concession and other measures.

NOTES

1. IEA Statistics, 2007.
2. EU member countries are Austria, Belgium, Cyprus, Czech republic, Denmark, Estonia, Finland, France, Germany, Greece, Hungry, Ireland, Italy, Latvia, Lithuania, Luxemburg, Malta, Poland, Portugal, Slovakia, Slovenia, Spain, Sweden, The Netherlands, United Kingdom.
3. The Koto Protocol, http://unfccc.int/essential_background/kyoto_protocol/items/1678.php.
4. Directive 2003/87/EC of the European Parliament and of the Council of 13 October 2003 establishing a scheme for greenhouse gas emission allowance trading within the Community and amending Council Directive 96/61/EC. Official Journal of the European Union of 25.10.2003 (L 275/32).
5. http://www.euractiv.com/en/sustainability/eu-emissions-trading-scheme/article-133629.
6. Communication from the Commission 'Energy for the Future: Renewable Sources of Energy: White Paper for a Community Strategy and Action Plan' COM(97) 599 final (26/11/1997), http://ec.europa.eu/energy/library/599fi_en.pdf as accessed on 17 October 2006.
7. Directive 2001/77/EC of the European Parliament and the Council of 27 September, 2001, on the promotion of electricity produced from renewable energy sources in the internal electricity market http://europa.eu.int/eur-lex/pri/en/oj/dat/2001/l_283/l_28320011027en00330040.pdf as accessed on 17 October 2006.
8. Communication from the Commission, The support of electricity from renewable energy sources, {SEC(2005) 1571} Brussels, 7.12.2005 COM(2005) 627 final http://ec.europa.eu/energy/res/biomass_action_plan/doc/2005_12_07_comm_biomass_electricity_en.pdf as accessed on 17 October, 2006.
9. European Commission, Directorate-General for Research: The State and Prospects of European Energy Research: Comparison of Commission, Member and Non-Member States' R&D Portfolios.
10. Communication from the Commission, The support of electricity from renewable energy sources, {SEC(2005) 1571} Brussels, 7.12.2005 COM(2005) 627 final.

11. Directive 2004/8/EC of the European Parliament and of the Council on the promotion of cogeneration based on useful heat demand in the internal energy markets on 11 February, 2004.
12. Directive 2003/30/EC of the European Parliament and of the Council of 8 May, 2003 on the promotion of the use of biofuels or other renewable fuels for transport.
13. Directive 2003/96/EC of the European Parliament and of the Council of 27 October 2003 on restructuring the community framework for the taxation of energy products and electricity.
14. Directive 2002/91/EC of 16 December, 2002 on the energy performance of buildings.
15. http://www.sustenergy.org/tpl/page.cfm?pageName=benchmarks as accessed on 1 November, 2006.
16. Statistisches Bundesamt, 2002.
17. Innovative energy technologies and climate policy in Germany, Katja Schumacher & Ronald D. Sands, Energy Policy 34 (2006) 3929-3941.
18. In 2000/2001, the German federal government reached an agreement to terminate use of nuclear energy and limit the operational lives of nuclear power plants to an average of 32 years, effectively phasing out nuclear energy by 2022. The amended version of the German Law on Nuclear Power entered into force on 27 April, 2002.
19. Innovative energy technologies and climate policy in Germany, Katja Schumacher & Ronald D. Sands, Energy Policy 34 (2006) 3929-3941.
20. Environmental Policy, Renewable Energy Sources in Figures—National And International Development, May 2006, Ministry for the Environment, Nature Conservation and Nuclear Safety (BMU), Germany.
21. *Ibid.*
22. For 2005 corresponds to: biodiesel: 1,800,000 tonnes approximately 2,045 million litres; vegetable oil: 196,000 tonnes approximately 213 million litres; bioethanol: 226,000 tonnes approximately 286 million litres.
23. Trends in Renewable Energies in 2005, Federal Ministry for the Environment, Nature Conservation and Nuclear Safety (BMU), Germany.
24. Trends in Energy; 2006, BMU, Germany.

25. Volkmar Lauber & Lutz Mez, Three Decades of Renewable Electricity Policies in Germany, Energy & Environment, vol. 15 no. 4 (2004), 599-623
26. Innovation through Research, 2005 Annual Report on Research Funding in the Renewable Energies Sector, German Federal Ministry of Environment, Nature Conservation and Nuclear Safety
27. Discussions with Mr. Joachim Nick- Leptin, BMU on 25 October 2006.
28. Practical Experiences with 300 MW wind power installed in Germany, Durstewitz, M. et al http://www.iset.uni-kassel.de/pls/w3isetdad/www_iset_page.show_ menu?p_lang=eng&p_owa_call=www_veroeff.show_veroeffdaten%3Fp_veroeff_nr=388%26p_lang=eng accessed on 2 November 2006.
29. Paul Runci, Renewable Energy Policy in Germany: An Overview and Assessment, Pacific Northwest National Laboratory, Technical Lab Report PNWD-3526 http://www.globalchange. umd.edu/energytrends/germany/ as accesed on 2 November, 2006.
30. Wilson Rickerson, *German Electricity Feed Law Policy Overview* http://www.ontario-sea.org/ARTs/Germany/RickersonGermanFeedLawPolicyOverview2002.doc accessed on 21 October, 2006.
31. Volkmar Lauber & Lutz Mez, Three Decades of Renewable Electricity Policies in Germany, Energy & Environment, vol. 15 no. 4 (2004), 599-623.
32. Gerhard Stryi-Hipp, The Effects of the German Renewable Energy Sources Act on Market, Technical and Industrial Development, Published at 19th European Photovoltaic Solar Energy Conference, Paris, 7-11 June, 2004.
33. Paul Runci, Renewable Energy Policy in Germany: An Overview and Assessment, Pacific Northwest National Laboratory Technical Lab Report PNWD-3526 http://www.globalchange. umd.edu/energytrends/germany/ as accessed on 21 October, 2006.
34. Non-official version of the Act from translation available at BMU website accessed on 30 October, 2006.
35. What Electricity from Renewable Energies Costs, A Publication from the German Federal Ministry for the Environment, Nature Conservation and Nuclear Safety, February 2006.
36. Volkmar Lauber & Lutz Mez, Three Decades of Renewable Electricity Policies in Germany, Energy & Environment vol. 15 no. 4 (2004), 599-623.

37. Details are available at http://www.kfw-foerderbank.de/EN_Home/Umweltschutz/index.jsp
http://www.kfw-foerderbank.de/EN_Home/Housing_Construction/index.jsp as accessed on 9 November, 2006.
38. Renewable Energies: Innovations for Future, Publication of the German Federal Ministry for the Environment, Nature Conservation and Nuclear Safety, April, 2006.
39. Biofuels in Germany and in the European Union: Technologies, Market Trends and Political Conditions by Mr Dieter Bockey Union zur Förderung von Oel- und Proteinpflanzen e.V. (UFOP) http://www.rlc.fao.org/prensa/activi/agroenergia/bockey.pdf accessed on 13 November, 2006.
40. Central Intelligence Agency, The World Factbook 2006, "United States," https://www.cia.gov/cia/publications/factbook/print/us.html, accessed August 25, 2006.
41. http://ocde.p4.siteinternet.com/publications/doifiles/012005061T018.xls and IEA Statistics 2006 accessed 1 September, 2006.
42. Source: Energy Information Administration, Electric Power Annual 2004, November 2005, http://www.eia.doe.gov/cneaf/electricity/epa/epa.pdf accessed August 25, 2006
43. http://www.eia.doe.gov/cneaf/electricity/epm/table1_1_a.html#_ftn1 as accessed on 4 September, 2006.
44. International Energy Agency, Renewable Energy: Market & Policy Trends.
45. http://www.doe.gov/energysources/hydrogen.htm accessed on 7 September, 2006.
46. http://genomicsgtl.energy.gov/biofuels/ accessed on 8 September, 2006.
47. http://www.doe.gov/news/3790.htm accessed on 7 September, 2006.
48. Data Base for State Incentives for Renewable Energy (DSIRE) http://www.dsireusa.org/summarytables/financial.cfm?&CurrentPageID=7&EE=0&RE=1 as accessed on 5 September, 2006.
49. Gielecki, Mark; Mayes, Fred; and Lawrence Prete, "Incentives, Mandates, and Government Programs for Promoting Renewable Energy," February 2001, Energy Information Administration, http://www.eia.doe.gov/cneaf/solar.renewables/rea_issues/incent.html, accessed 5 September, 2006.

50. Gielecki, Mark; Mayes, Fred; and Lawrence Prete, "Incentives, Mandates, and Government Programs for Promoting Renewable Energy," February 2001, Energy Information Administration, http://www.eia.doe.gov/cneaf/solar.renewables/rea_issues/incent.html#N_1 accessed 5 September, 2006.
51. In 1995, FERC defined avoided costs as, "...the incremental costs to an electric utility of electric energy or capacity or both which, but for the purchase from the qualifying facility or qualifying facilities, such utility would generate itself or purchase from another source." FERC, "18 CFR Part 292 Regulations Under Sections 201 and 210 of the Public Utility Regulatory Policies Act of 1978 With Regard to Small Power Production and Cogeneration," accessed 5 September, 2006.
52. For more information on how these incentives changed over time, please see: Gielecki, Mark; Mayes, Fred; and Lawrence Prete, "Incentives, Mandates, and Government Programs for Promoting Renewable Energy," February 2001, Energy Information Administration, http://www.eia.doe.gov/cneaf/solar.renewables/rea_issues/incent.html accessed 5 September 2006.
53. http://www.ucsusa.org/clean_energy/clean_energy_policies/production-tax-credit-for-renewable-energy.html accessed on 6 September, 2006.
54. http://www.eia.doe.gov/cneaf/solar.renewables/page/epact/epact.pdf accessed on 7 September, 2006.
55. (a) http://www.dsireusa.org/documents/SummaryMaps/RPS_Map.ppt assessed on 6 September 2006; (b) Energy Information Administration, http://www.eia.doe.gov/oiaf/analysispaper/rps/index.html, accessed on 8 September, 2006; and (c) Energy Information Administration, Annual energy outlook 2006 with projections to 2030, report no. DOE/EIA-0383(2006), Washington, DC: U.S. Department of Energy, Energy Information Administration, February, 2006.
56. Some States limit eligibility to photovoltaic systems while others permit all small renewable facilities to participate.
57. DSIRE, "Net Metering Rules," http://www.dsireusa.org/library/includes/seeallincentivetype.cfm?type=Net&currentpageid=7&back=regtab&EE=1&RE=1 accessed 6 September, 2006.
58. http://www.dsireusa.org/summarytables/reg1.cfm?&CurrentPageID=7&EE=1&RE=1 accessed on 6 September, 2004.

59. DSIRE, "Net Metering Rules," http://www.dsireusa.org/library/includes/seeallincentivetype.cfm?type=Net&currentpageid=7&back=regtab&EE=1&RE=1 accessed 6 September, 2006.
60. http://www.energy.ca.gov/electricity/gross_system_power.html accessed on 8 September, 2006.
61. California Energy Commission, "Net System Power: A small Share of California's Power mix in 2005", April 2006, Report no CEC-300-2006-009-F http://www.energy.ca.gov/2006publications/CEC-300-2006-009/CEC-300-2006-009-F.PDF accessed on 8 September, 2006.
62. California's 2005 Integrated Energy Policy Report, http://www.energy.ca.gov/2005_energypolicy/index.html accessed on 12 September, 2006.
63. For a comprehensive list of current renewable energy incentives in California, see, http://www.dsireusa.org/library/includes/statesearch2.cfm?State=CA&back=fintab&CurrentPageID=7&Search=TableState&EE=1&RE=1, accessed on 11 September, 2006.
64. Dodge, Darrell M, "Illustrated History of Wind Power Development," http://telosnet.com/wind/recent.html accessed on 11 September, 2006.
65. DSIRE, "Net Metering Rules—California http://www.dsireusa.org/library/includesincentivesearch.cfm?Incentive_Code=CA02R&Search=TableType&type=Net&CurrentPageID=7 accessed on 12 September, 2006.
66. *Ibid*, accessed on 12 September, 2006.
67. *Ibid*, accessed on 12 September, 2006
68. DSIRE, California's Renewable Portfolio Standard (RPS), http://www.dsireusa.org/library/includes/tabsrch.cfm?state=CA&type=RPS&back=regtab&Sector=S&CurrentPageID=7&EE =0&RE=0 assessed on 12 September, 2006.
69. California Energy Commission, Integrated Energy Policy Report (CEC-100-2005-007 CMF), November 2005.
70. History of Renewable Programme, California Energy Commission, http://www.energy.ca.gov/renewables/history.html accessed on 12 September, 2006.
71. California Energy Commission, http://www.newsolarhomes.ca.gov/background.html accessed on 12 September, 2006.
72. California Energy Commission, http://www.consumerenergycenter.org/erprebate/program.html accessed on 12 September, 2006.

73. Energy Information Administration, Annual Energy Outlook 2006 with projections to 2030, report no. DOE/EIA-0383(2006), Washington, DC: U.S. Department of Energy, Energy Information Administration, February 2006
74. https://www.cia.gov/cia/publications/factbook/print/ja.html as accessed on 21 November, 2006.
75. http://www.iea.org/Textbase/stats/oecdcountryresults.asp?oecd=Japan as accessed on 21 November, 2006.
76. REN 21, Renewables, Global Status Report, 2006 Update
77. http://www.iea.org/Textbase/stats/electricityoecd.asp?oecd=Japan&COUNTRY_LONG_NAME=Japan as accessed on 21 November, 2006.
78. http://www.iea.org/textbase/nppdf/free/2004/renewable1.pdf as accessed on 21 November, 2006.
79. *Ibid.*
80. *Ibid.*
81. IEEJ as accessed on 21 November, 2006.
82. http://www.oja-services.nl/iea-pvps/ar04/index.htm as accessed on 22 November, 2006.
83. http://www.nedo.go.jp/english/archives/161027/pv2030roadmap.pdf as accessed on 22 November, 2006.
84. IEA and http://www.gwec.net/uploads/media/06-02_PR_Global_Statistics_2005.pdf as accessed on 22 November, 2006.
85. http://www.enecho.meti.go.jp/english/newnationalenergystrategy2006.pdf as accessed on 22 November 2006.
86. http://www.iea.org/Textbase/stats/balancetable.asp?COUNTRY_CODE=CN as accessed on 22 February, 2007.
87. http://www.nrel.gov/international/china/pdfs/35788_township.pdf as accessed on 5 December, 2006.
88. http://www.iea.org/Textbase/stats/electricitydata.asp?COUNTRY_CODE=CN as accessed on 22 February, 2007.
89. Based on http://www.martinot.info/china.htm accessed on 23 November, 2006.
90. http://www.nrel.gov/docs/fy04osti/35790.pdf accessed on 5 December, 2006.
91. Based on http://www.martinot.info/china.htm accessed on 23 November, 2006.
92. http://www.nrel.gov/international/china/pdfs/35788_township.pdf accessed on 5 December, 2006.

93. From China: Prospect for Renewable Energy Development by Li Junfeng1, Shi Jinli and Ma Lingjuan http://www.hm-treasury.gov.uk/media/999/B2/Final_Draft_China_Mitigation_Renewables_Sector_Research.pdf accessed on 5 December, 2006.
94. IEA: http://www.iea.org/textbase/pamsdb/grcountry.aspx?country=China accessed on 22 February, 2007.
95. The authorized text of the Chinese Renewable Energy Law may be seen at http://china.lbl.gov/publications/re-law-english.pdf.

Chapter 4

Renewable Energy in India: Status and Prospects

Energy services always remained in focus of the Indian successive governments and concerted efforts were made to improve the energy infrastructure within the country. The economic policy changes undertaken since early 1990s with greater participation of private sector and deregulation of infrastructure sectors has resulted in higher growth rates of the Indian economy and thereby an increased demand for energy. However, the growth of energy sector could not match with the growth in economy and the energy sector has been viewed as a substantial bottleneck to India's industrial growth that in turn is seen as critical to stimulating the country's economic and social development.

4.2 India Energy Sector

The country has seen an expansion in total energy use during the last five decades, with a shift from non-commercial to commercial sources of energy. Accordingly, the production of commercial sources of energy has increased significantly. As such there has been over tenfold increase in India's commercial energy supply. In 2003-04, the total commercial energy supply (TCES) was 327 million tonnes of oil equivalent (mtoe) as compared to about 26 mtoe in 1953-54.[1] One of the important feature of Indian energy mix is high share of traditional sources (mainly combustible renewables and waste) of energy. Although the country's dependence on traditional fuels has reduced substantially over the years, the Planning Commission of India has estimated that even in 2005-06 over 30 per cent of India's energy supply is derived from traditional fuels. These sources accounted for about three-quarters of the country's energy supply in 1953-54.

Taking traditional sources into account, India's total primary energy supply raise around five times between 1953 and 2004. Despite this, India is one of the lowest in the world in terms of per capita energy consumption. In per capita energy consumption terms, it is 31 per cent of the global average, 7 per cent of USA, 12 per cent of Germany and 48 per cent of China's per capita total primary energy consumption.

Coal and oil account for most of India's commercial energy. Being the most abundant fossil fuel in India till date, coal continues to be one of the most important sources for meeting domestic energy needs. It accounts for 55 per cent of the country's total energy supplies. The share of oil and gas in the primary commercial energy consumption of the country has increased from 9.4 per cent in 1954 to 44.9 per cent (36.0 per cent for oil and 8.9 per cent for natural gas) in 2003-04. India is not self-sufficient in energy as its imports accounted for over 25 per cent of the TCES in 2003-04.[2]

Table 4.1 Total Primary Energy Supply 2003-04

(*in Mtoe*)

Source	*Total*
Coal & Lignite	167.00
Oil & Products	119.00
Gas	29.00
Electricity	12.00
TOTAL COMMERCIAL ENERGY	327.00
Non Commercial	143.00
TOTAL ENERGY	470.00

Power sector alone consumes 75 per cent of the coal produced in the country.[3] Around half of the total petroleum consumption in the country is towards the transport sector. The share of coal in the total final energy consumption has been steadily declining while that of oil and gas has been increasing. As such, industry and transport are the two largest consumers of commercial energy, accounting for around 50 per cent and 24 per cent, respectively, of the total in 2003-04.

On the electric power front, India accounted for 3.1 per cent of global electricity consumption in 2004.[4] The Indian power planning focus on extension of the centralized grid throughout the country. India had an installed capacity of 128432 MW as on 16 February, 2007.[5] Of this 84,440 MW is accounted for by thermal power plants, 33942 MW of large hydro plants, 3900MW of nuclear and the balance 6191 MW from renewable sources. 55.4 per cent of this total capacity is within state electricity boards and electricity departments, 33.7 per cent in the central sector and 10.9 per cent in the private sector.[6] The gross generation by utilities was 594 billion kWh in 2004-05 of which thermal generation accounted for 82 per cent, hydro 14 per cent, and nuclear 3 per cent. In addition, the capacity in captive power plants of 1MW and above was estimated to be about 19,000 MW, with around 2350 plants as on 31st March 2005.[7]

The major consumers of electricity from utilities are industry (35.6 per cent), agriculture/irrigation (22.9 per cent), residential (24.77 per cent), commercial (8.13 per cent), and railways (2.46 per cent).[8] Average per capita consumption has grown from 15.6 kWh in 1950 to about 612.50 kWh for the year 2004-05 which is still much lower than consumption standards prevailing in developed economies.

The capacity addition has not been able to keep pace with the increasing demand for electricity and there has been persistent energy and peak shortages in the country. During April 2006—January 2007 there was a power shortage of 9.3 per cent and the deficit in peak energy demand was 13.9 per cent. This apart, a significant portion of the electricity generated remains uncounted. During 2004-05 the electricity lost in transformation, transmission and distribution system and electricity unaccounted was around 35.03 per cent of total available energy.[9] The Ministry of Power, Government of India has planned to develop over 62,000MW of capacity during the period 2007-2012. However, in view of the current economic growth trends the planned capacity is unlikely to meet the electricity demand by the year. This apart, according to 2001 census 12.5 per cent urban households and 56.5 per cent rural household were still un-electrified.[10] Out of 593732 inhabited villages, 119570 villages were not having grid extension as on 15 May, 2006.[11] As such the present power sector is characterised by capacity shortages, frequent blackouts and poor reliability.

One of the major challenges is that of providing access to modern energy sources to a large proportion of the country's population. Around 86 per cent of rural households and more than 20 per cent of the urban households still rely primarily on traditional fuels such as firewood, wood chips or dung cakes to meet their cooking needs. The use of traditional fuels entails high opportunity cost by the way of the health problems arising from indoor air-pollution caused by their extensive use and the number of hours of productive employment or schooling foregone in their collection. Only 5 per cent and 2.7 per cent of rural households use LPG and kerosene respectively as a primary cooking fuel where as 44 per cent and 22 per cent of urban households uses LPG and kerosene respectively.[12]

Given the importance of the energy sector for advancing India's human and economic condition, a sizable fraction of government expenditures have been targeted towards its development. For example, from the 4th five-year plan onwards, the energy sector has accounted for more than 20 per cent of the total expenditures/outlay in the Plans. In the 10th Plan (2002-2007) it was allocated over 24 per cent of the public sector outlay.[13]

4.3 Energy Projections and Policies

The Indian Energy sector being central to economic development has been the focus of attention of the respective governments.

Since 1990s this sector like the rest of India's economy, is undergoing large changes.

Projections made by the Integrated Energy Policy Committee of the Planning Commission has estimated that in order to meet the projected GDP growth of 8 per cent per annum by 2031-2032, the demand for primary energy will increase to 1836 Mtoe representing almost a 4 fold increase since 2003-04. With regard to commercial energy the requirement would be around 1651 Mtoe, which is around 5 fold increase since the year 2003-04. The non-commercial energy requirement is expected to increase only 1.3 times during the corresponding period. It has also estimated that by the period the power generation capacity must increase to nearly 800000 MW.

In the event of achieving the aforesaid capacities, the per-capita primary energy consumption will be expected to be almost 1250 kgoe, which will be still much lower than the current international average of about 1770 kgoe/capita and the levels in industrialized countries of 4730 kgoe/capita.

The Committee in its report has recommended a number of policy changes and technology development initiatives for achieving energy security. These include ensuring adequate supply of coal of consistent quality and availability of gas for power generation, continuing power sector reforms, reduction in cost of power, rationalisation of fuel prices, energy efficiency and demand side management, augmenting of resources for increased energy security, augmenting energy availability through overseas arrangements, realizing hydro and nuclear potential and developing renewable resources alongwith appropriate policy measures.

Development of indigenous resources such as coal, oil, gas, hydro, nuclear and renewables and additions to downstream processing capabilities for these energy sources would be important in meeting the challenge of the future commercial energy demand. Alongside, significant expansion of the infrastructure in terms of roads, rails, ports and rural connectivity would be required. Availability of financial resources from government as also private sector including foreign funds for achieving the desired level of infrastructure development would be crucial.

The government has already realized the sense of urgency and accordingly reform process has started in energy sector which include new electricity act, energy conservation act, rural electrification policies and liberal policies for foreign investment in the energy sector. This apart, restructuring and strategic privatization of public sector institutions has been contemplated in order to improve the overall system efficiency. The issues are really complicated and multifaceted as expansion of energy services has direct bearing upon national financial balance as substantial portion of export earnings are used for meeting energy import needs even at the present levels and with many fold increase in energy imports likely in future, the government would need to work out an export strategy which meets growing requirement of imported energy without adversely affecting the national balance of payments. In view of limited indigenous hydrocarbon reserves available with the country increasing energy imports is apparent.[14]

4.3.1 Coal

Coal mining in India dates back to the 18th century and India now ranks 3rd amongst the coal producing countries in the world. It accounted for over 50 per cent of the country's energy need in 2004-05. At the present rate of coal extraction, the proven reserves are likely to last around 90 years. Power utilities are the major consumers of coal in the country, accounting for around 78 per cent of the total coal offtake. In addition, other industries like steel, cement, fertilizers, chemicals, paper besides thousands of medium and small-scale industries are also dependent on coal for their process heat and energy requirements.

It has been estimated that at a modest growth rate of production of 5.5 per cent per year the production of coal and lignite by 2031-32 will be around 1400 Mt and import needs for the various scenarios with an 8 per cent GDP growth rate an import dependence for energy in 2031-32 could be as low as 29 per cent and as high as 59 per cent. As such augmenting coal resources and production, while minimizing environmental impact, is thus accorded high priority in government planning.[15]

Table 4.2 Ranges of Commercial Energy Requirement, Domestic Production and Imports for 8 per cent Growth for Year 2031-32

Fuel	*Range of Requirement in Scenarios*	*Assumed Domestic Production*	*Range of Imports*	*Import (Per cent)*
Oil (Mt)	350–486	35	315–451	90–93
Natural Gas (Mtoe) including Coal Bed Methane	100–197	100	0-97	0-49
Coal (Mtoe)	632-1022	560	72-462	11-45
Total Commercial Primary Energy Supply	1351-1702	—	387-1010	29-59

4.3.2 Petroleum and Natural Gas

The country currently imports over 70 per cent of its oil and this share is expected to exceed 90 per cent by 2031-32. The production of crude oil in the country has increased from 6.82 Mt in 1970-71 to 33.38 Mt in 2003-04. The quantity of crude oil imported increased from 11.66 Mt during 1970-71 to 87 Mt by 2004-05.[16]

In absolute terms oil consumption is estimated to increase 3-4 times to the present level by 2031-32. However, long-term growth in demand of oil depends upon a number of factors such as economic growth (GDP), elasticity of demand for petroleum products, including oil with respect to GDP growth, relative price levels of substitute products including bio-fuels, growth of road infrastructure, the price of oil, the future efficiency of vehicles and the growth of alternate modes of transport. Naphtha demand is dependent on the growth plans for fertilizer and petrochemicals and its price relative to the price and availability of natural gas.

Gas import was started in 2004 and is projected to reach an import dependency of around 50 per cent under the scenario of minimum domestic production and maximum consumption. Gas constituted around 9 per cent of Indian commercial energy mix in 2003-04 and is likely to increase to around 12 per cent by 2031-32 under 8 per cent GDP growth scenario. In terms of volumes the consumption is likely to be 6 fold from that of 2003-04 level of 29 Mtoe.

4.3.3 Atomic Energy

Atomic energy accounts for around 3.0 per cent of the country's installed capacity and electricity generation. However, in the long-term nuclear power development is considered to be a critical element of any strategy to avoid excessive dependence on energy imports in the coming decades. According to the Energy Policy Committee, Planning Commission projections under an optimistic scenario around 63000 MW electric generating capacity can emanate from nuclear power by 2030. However, such capacity addition would require successful demonstration of Fast Breeder Reactor Technology (FBR), perfection in light water reactor technology and also augmenting Uranium availability through new mines and developing the Advanced Heavy Water Reactor for utilising Thorium by 2020. In addition, it is assumed that India will be able to import 8,000 MW of Light Water Reactors with fuel over the next ten years. As such Available Uranium supply can fuel only 10,000 MW of the Pressurized Heavy Water Reactors (PHWR).

4.3.4 Electric Power

As per Integrated Energy Policy Report around 780,000 MW generating capacity would be required to meet the electricity needs for 8 per cent per annum GDP growth by 2031-32. The contribution of fossil fuel, mainly coal would remain around 78 per cent of total electricity generated by the time.

Table 4.3 Projections for Energy Requirements

Year	*Total Energy Requirement (billion kWh)*	*Projected Peak Demand (GW)*	*Installed Capacity (GW)*
2011-12	1097	158	220
2016-17	1524	226	306
2021-22	2118	323	425
2026-27	2866	437	575
2031-32	3880	592	778

There is another estimate from Sixteenth Electric Power Survey of India of CEA which suggests that by 2016-17 there would be requirement of 1318 billion kWh and the peak load requirement would be around 212,700 MW.

Table 4.4 Region-wise Demand Projections for Public Utilities

Region	*Energy requirement (MU)*		*Peak load (MW)*	
	2011-12	*2016-17*	*2011-12*	*2016-17*
Northern	3,08,528	4,29,480	49,674	69,178
Western	2,99,075	3,95,859	46,825	61,966
Southern	2,62,718	3,54,599	42,061	56,883
Eastern	90,396	1,17,248	15,664	20,416
North-Eastern	14,061	20,756	2,789	4,134
Islands				
Andaman & Nicobar	374	591	77	122
Lakshadweep	70	111	17	26
TOTAL	975,222	1,318,644	157,107	212,725

While electricity supply has reached all towns and 85 per cent of the villages in the country, about 57 per cent of rural and 12 per cent urban households did not have electricity in 2000. Recognizing that electricity is one of the key drivers for poverty alleviation, the Electricity Act, 2003 requires that the government shall endeavour to supply electricity to all areas including villages and hamlets by 2012. The Government has also created the

Central Electricity Regulatory Commission under the Electricity Regulatory Commission Act of 1998 and some states have set up counterpart bodies. The Electricity Bill 2003, seeks to further address and facilitate power sector reform. The National Electricity Policy notified in January 2005 has set a goal of access to electricity available for all households in next five years and per capita availability of electricity to be increased to over 1000 units by 2012 alogwith a minimum lifeline consumption of 1 unit/household/day as a merit good. In order to meet this target, a comprehensive national scheme with an outlay of Rs. 160 billion was launched in April 2005. The Rajiv Gandhi Grameen Vidyutikaran Yojana (RGGVY) aims to electrify 1,25,000 unelectrified villages; connect all the estimated 23.4 million unelectrified households below the poverty line with 90 per cent subsidy on connection costs; and augment the backbone network in the 0.46 million already electrified households. The 54.6 million households above the poverty line, which are currently unelectrified, are expected to get electricity connections without support.[17] Simultaneously, the government is promoting the use of renewables to cater to villages located in remote and geographically isolated areas, where extension of the existing electricity grid is not economically viable.

The Programme for Electrification of Remote Villages and Remote Hamlets through renewable sources is designed to provide decentralized power generation facilities in the remote areas. Under this programme, up to 90 per cent financial assistance is provided by the government, while ensuring local participation for the sustainability of the programmes. As of 30 June, 2006, 2237 remote villages and 594 remote hamlets have been electrified under this programme.

India has been making ambitious efforts to increase its electricity generation capacity. Due to the paucity of the resources available to the Central and State PSUs and the state electricity boards (SEBs) for developing the additions needed, the Indian Government opened up India's power sector in 1991, making a series of major changes in its policies at that time, and subsequently, in order to attract private investment. However, still the share of private sector in total electricity generation is comparatively small at around 11 per cent.

4.4 New and Renewable Energy

The history of evolution of renewable energy technologies in the country has been interesting. The Indian scientists had initiated research and development efforts in certain areas like biogas generation from cow-dung even before independence in 1930 at the Indian Agricultural Research Institute (IARI) Pusa, New Delhi. The activities relating to harnessing other renewable energy sources were initiated by large group of National and Regional Research Laboratories under the Council of Scientific and Industrial Research (CSIR) just after independence in 1950s.

Major institutions engaged in solar development included National Physical Laboratory (NPL), Central Salt and Marine Chemicals Research Institute, Central Building Research Institute, Central Arid Zone Research Institute, Indian Institute of Technology at Delhi and Chennai, and Aurovile Centre for Environmental Studies. NPL developed a corrugated type solar water heater whereas the solar water heater developed by CBRI was handed over to NRDC in 1968 for commercial application. Solid State Physics Laboratory, Ministry of Defence intiated research and development activity on silicon solar cells around 1966 and the process know-how for fabrication of silicon solar cells was transferred to NRDC in 1971.

In wind energy area under the project "Utilization of Wind Power" National Aeronautical Laboratory (NAL) commenced work in 1952 and for assessing potential of wind energy in India more than 200 stations scattered all over India were considered. Two types of water pumping wind mills were developed and field tested upto 1966. Designing of 250 watt, 1 KW and 5 KW wind electric generators were taken up in collaboration with an UN expert from the former USSR.

IARI continued to work on biogas technology development. Since 1962 Khadi and Village Industries Commission (KVIC) started setting up floating gas holder type biogas plants.

In 1972, the National Committee on Science and Technology (NCST) under the Department of Science and Technology, Government of India constituted among others a panel on fuel and power which further constituted several committees to develop specific plans and proposal for developing new and renewable energy in the country. In its reports submitted to NCST these committees gave specific recommendations alongwith concrete proposals in the areas of solar energy, wind energy, biogas, fuel cells, tidal power and geothermal energy.[18]

In 1976 the Department of Science and Technology created a new energy source division to develop renewable energy in a focused manner. In January 1978 India hosted International Solar Energy Society Congress at New Delhi where detailed action plan for technology development and status of solar technologies were discussed. The political wisdom to act was reflected in the India's active participation led by the then Prime Minister Smt Indira Gandhi in the UN Conference on New and Renewable Sources of Energy held at Nairobi in 1981. This was the first major attempt of the international community to develop a strategy for the use of alternative fuels. All these efforts ultimately culminated in constitution of a Commission for Additional Sources of Energy (CASE) under the Department of Science & Technology for the purpose of developing and promoting new and renewable sources (NRSE) of energy in March 1981. In the resolution setting up CASE, the Government of India observed that it would be "a mechanism, free from all non-essential restrictions and needlessly inelastic rules, which will have responsibility in the entire field of development of NRSE, coordinating and intensifying R&D activities in the area and ensuring the implementation of the Government's Policies in regard to all matters concerning NRSE." The Commission was required to recommend the Government incentive measures for commercial use of NRSE technologies by industries and also to develop and coordinate with other agencies, schemes oriented to the needs of rural India.[19] Initially CASE functioned in the Department of Science and Technology. Recognizing the importance of developing NRSE in larger framework, the Government of India created the Department of Non-Conventional Energy Sources (DNES) on 6th September 1982. The thrust areas identified included research, development, demonstration and dissemination of renewable energy technologies for providing energy services in the rural areas and also for meeting energy needs of the country through renewable sources. Later in 1992, DNES was upgraded into an independent Ministry of Non-Conventional Energy Sources (MNES), which was further renamed as Ministry of New and Renewable Energy (MNRE) on 20 October, 2006.

The Scientific Advisory Committee of the Cabinet (SACC) has included MNRE in the list of Scientific Departments/Agencies. However, MNRE is not merely a Scientific Ministry, its responsibilities also include development, extension and promotion of all new and renewable sources of energy. In addition, its programmes have a social and environmental dimension that addresses cross-cutting issues of socio-economic development, green house gases emission reduction, rural health, sanitation and also livelihood. It also provides linkages with commerce and industry.

In 1987 a separate financing institution called the Indian Renewable Energy Development Agency (IREDA) was set up as a public sector undertaking for providing institutional finance exclusively in the field of renewables and energy efficiency. IREDA entered the market when lending to the renewables sector was considered a high risk and low profit business. Over the years, IREDA has paved the way for broadening and deepening the market for renewables. In the initial years, it received bilateral and multilateral credits. Over the years other financing institutions also started providing financial assistance for renewable energy projects and consequently the market share of IREDA financed projects has at present come to around 20 per cent.

4.4.1 Renewable Energy in Indian Planning Process

Renewable Energy sector got specific attention in the planning process since sixth five-year plan with specific budget allocation. The programmes during 1980-85 mainly focused on research and development and gathering of field information on potential of renewables and their commercial and technical viability. In terms of technology development biogas programmes had got certain technological maturity and other renewable technologies such as solar, wind, etc. were yet to establish their viability for large scale application. Therefore, in addition to research the emphasis was on setting up demonstration projects.

The main element of the 7th plan strategy was to encourage deployment of renewable energy wherever technically viable and support intensive research and development of indigenous technology. Additional emphasis was made on improving access to and availability of renewable technologies. During the period states were mobilized to set up dedicated nodal agencies to serve as focal points for renewable energy activities and undertake the development and dissemination of renewable energy systems in the respective states.

During the 8th plan period under economic liberalization process, the programme emphasis shifted from purely subsidy-driven dissemination programs to technology promotion through commercial route. Whereas rural oriented renewable energy programmes continued with direct subsidies, market orientation strategy for promotion of renewables was through fiscal incentives and elimination/reduction of direct subsidies, encouragement to private entrepreneurs for investing in renewable energy technologies through measures like accelerated depreciation and reduced import duties.

The strategy for the 9th plan document highlighted the need for concerted efforts for developing environment friendly technologies. In regards to renewable programmes it suggested shift from capital subsidy to interest subsidy and gradual movement towards private entrepreneurship and community participation. During plan period the specific thrust was towards orienting the structure of the ongoing programmes and structuring socially oriented programme aiming at reduction in subsidy to minimum level and also defining time frame to phase them out.

The 10th plan suggested introduction of a suitable policy framework that provides for remunerative returns and encourages private investment in renewables. Accordingly, 10th plan strategy was of continuing budgetary support for research and development, demonstration and rural oriented programmes and promoting private investments through fiscal and promotional policy support.

The 11th Plan approach paper has observed that "Renewables such as wind power can play a useful role in the 11th Plan. These can be set up in a short time and though they have a low load factor, with a good transmission system, can be balanced by hydro power and can contribute to meeting peak demands." It also proposes to move towards a more transparent policy framework that treats different sources of energy in a similar fashion. In this context it has suggested to "Treat environmental externalities uniformly under the 'polluter pays' principle. Thus renewables should be given appropriate incentives." [20]

4.4.1.1 Integrated Energy Policy

The Prime Minister and the Deputy Chairman, Planning Commission, Government of India, took the decision for an effective and comprehensive energy policy as an urgent imperative in the year 2004. An expert committee was constituted under the leadership of Dr. Kirit Parikh, Member, Planning Commission to prepare an integrated energy policy linked with sustainable development that covers all sources of energy and addresses all aspects including energy security, access and availability, affordability and pricing, efficiency and environment. The committee was constituted on 12 August, 2004. The draft integrated energy policy was circulated in December 2005 and the final policy was notified in August 2006. The Committee recommended that renewable energy needs special policies to encourage them in a time bound manner not just by putting some outpays in the planning process. The recommendations of the Committee included:

(a) Phase out capital subsidies, which only encourage investment without ensuing outcome, by the end of the 10th Plan linked to creation of renewable grid power capacity;
(b) Power regulators must seek alternative incentive structures that encourage utilities to integrate wind, small hydro, cogeneration, etc., into their systems. All incentives must be linked to energy generated as opposed to capacity created; and
(c) Respective power regulators should mandate feed-in laws for renewable energy, where appropriate, as provided under the Electricity Act and as are mandated in many countries.

The specific policies to promote various renewable energy technologies as recommended by Committee included:

(a) *Mini Hydro*: A detailed survey should be carried out to identify potential sites. Identified sites should be auctioned. For plants which are not connected to grid bid for lowest tariff with a pre-specified premium in the form of Tradable Tax Rebate Certificates (TTRC) should be invited. For village level plants, the entrepreneurs should be encouraged to supply power to meet other requirements such as agro processing and milling. If the plant can feed into a grid, the grid should be required to accept power at the going time of day tariff, and the plant site should be auctioned off for minimum premium in the form of TTRC linked to output. The responsibility for investments for connecting to the grid should be fixed in advance before the bidding.

(b) *Wind Power*: For wind power, site selection is freer than hydro-power and wind plants can be set-up on private land. Thus there may be need to auction only sites on public property. The same two types of auctions may be followed as described above for hydro-power plants.

(c) *Fuel-wood Plantation*: Cooperatives should be encouraged and facilitated to grow tree plantations in villages. Cooperatives which are open to all members of the community and which are non-discriminatory should be given government land on long-term lease. Women should be encouraged to set-up and manage such plantations so that the time they now spend in gathering fuel can be spent productively in a way that empowers them. They should also be provided finance. If organized and managed properly, such plantations are economic and successful. Field based NGOs could also be involved in this activity. To encourage large-scale plantations, contract farming should be facilitated.

(d) *Electricity from Wood Gasification*: This can provide electricity based on gasification of wood and can be very useful especially in remote villages. The same set of policies, indicated for micro hydel and wind power plants should be followed here.

(e) *Bio Gas Plants*: The real potential of bio-gas is in community level plants. To encourage private or community entrepreneurs to set these up, they need to be provided land and finance. Also to have the willing participation of all the cattle owners in the community requires an appropriate operating strategy. The essential policy required is provision of land and finance.

(f) *Solar Thermal Power Plants*: In order to encourage entrepreneurs to invest in solar thermal plants, a higher premium of feed-in tariff may be given. The higher premium can be justified given the higher risk and may be available to only the first 5000 MW of solar thermal plants.

(g) *Solar Photovoltaics*: In view of its very high ultimate potential, incentive to commercialise and lower the cost for solar photovoltaic power plants may be provided through a higher feed-in tariff, again for the first 5000 MW of installed capacity.

The Committee also made specific recommendations for renewable energy technology development. These are summed up in the chapter on research and development.

4.4.2 *Renewable Energy Potential in India*

There have been a number of estimates about renewable energy potential in the country. These are generic estimates and in real terms renewable energy potential depends on various parameters. For all practical purposes the potential could be categorized in different categories namely, theoretical, technological, economic, conversion and sustainable potential. The theoretical potential identifies the physical upper limit of the energy available from a certain source. The conversion potential is derived from the theoretical potential and the annual efficiency of the respective conversion technology. The conversion potential as such is not a strictly defined value, since the efficiency of a particular technology depends on technological progress. The technological potential could be derived from the conversion potential, taking account of additional restrictions regarding the area that is realistically available for energy generation. Like the conversion potential, the technological potential of the different energy sources is therefore not a strictly defined value, but depends on numerous boundary conditions and assumptions. Further, at a point of time what matters is what proportion of technological potential could be utilized economically, based on economic boundary conditions (at a certain time). For biomass, for example, those quantities are included that can be exploited economically in competition with other products and land uses. The economic boundary conditions can be influenced significantly, particularly through political measures. The Renewable Energy Potential is given in Table 4.5.

Table 4.5 Renewable Energy Potential

Source	*Potential*
Wind Energy	45,000 MW
Small hydro (upto 25 MW)	15,000 MW
Biomass Power (Agro residue)	17,000 MW
Cogeneration baggase	5,000 MW
Waste to Energy	2,700 MW
Solar Energy	20MW/sq.km
Solar Water Heater	140 million square meter collector area
Family type Biogas Plants	12 million
Improved cook stove	120 million

4.4.2.1 Wind Power

Wind power is characterized according to wind power class on a classification scale. There are 490 dedicated wind-monitoring stations covering 20 states and 3 union territories. The assessments made so far indicate that India has low wind regime and most of potential is in the peninsular India. As per the assessment made by C-WET the wind power potential on macro level based data collected from 10 states with wind farm requiring @12 ha/MW and considering only 1 per cent of land availability is around 46092 MW.[21]

Most of the sites are in low wind area and as per wind speed classification around 90 per cent potential falls in under class 2 wind regime which are considered poorest among all the sites.

Table 4.6 Estimated Wind Electricity Potential

Wind Power Class	*Wind characteristics at 50 m above ground*		*Potential (in MW)*
	Wind speed m/s	*Wind power w/m²*	
I	< 5.6	< 200	—
II	5.6 — 6.4	200 — 300	43466
III	6.4 —7.0	300 — 400	4683
IV	7.0 — 7.5	400 — 500	395
V	7.5 — 8.0	500 — 600	17
VI	8.0 — onwards	600 — onwards	—

As the potential is sensitive to land availability, in a scenario of higher land availability, the potential would accordingly increase. Further, off-shore wind potential is yet to be mapped and estimated, which in a longer time horizon could be feasible.

4.4.2.2 Small Hydro

In the Indian context small hydro power (SHP) means plants with capacity uto 25 MW capacity. Gross estimate indicates that there exists a potential of around 15,000 MW of Small Hydro in the country and around 4,404 sites with an aggregate capacity of 10,477 MW have been identified. Most of this potential is in the hilly Himalayan region and in Maharastra, Kerala and Karanataka.

4.4.2.3 Solar Energy

India is located in the equatorial sun belt of the earth, thereby receiving abundant radiant energy from the sun. In most parts of India, clear sunny weather is experienced over 300 days a year. The annual average solar energy falling on one square meter of area is in the range of 4-7 kilowatt-hour and the average solar insolation in the country has been estimated at 6 kWh/meter^2/day, which is comparable with radiation received in the tropical and sub-tropical regions. The highest annual global radiation is received in Rajasthan and northern Gujarat. The solar energy potential can be exploited by many direct thermal applications such as for cooking, heating, power generation or in photovoltaic cells that directly convert sunlight to electricity.

As such most parts of India has solar radiation considered suitable for large scale solar power generation projects. Apart from the radiation, the land availability is an important requisite for utilization of solar energy. Considering that India is a very densely populated country, agricultural land and forests cannot be considered for large scale solar plants. Thus, only a portion of waste lands which has been estimated at around 640,000 square kilometer can be considered for such plants. Most of the wastelands are located in the central part and the north-western part of India where the solar radiation is highest. Rajasthan and Gujarat with around 150,000 square kilometer of wastelands, most of it deserts, and annual global solar radiations of 2100 to more than 2350 kWh/m^2, can be considered most appropriate locations. About 10 per cent of this waste land would *i.e.* around 15000 square kilometer could be conveniently be made available for solar power projects. In order to understand potential let us undertake a hypothetical exercise. During 2004-05, around 386 TWh electricity was sold to consumers in the country.[22] In case 40 sqkm area in Rajasthan desert is covered by Solar PV cells, with present conversion efficiency of commercially available photovoltaic cells of around 15 per cent, 432 TWh electricity could be generated in a year if systems work 6 hrs per day for 300 days a year. However, if system efficiency could be raised from the current level of 15 per cent assumed herein, the land mass area required for the purpose would reduce proportionately. This indicates that land availability will not be the limiting factor in future.

Another major solar energy application is for thermal applications such as water heating, drying, cooking, etc. MNRE has estimated solar thermal potential at around 140 million sq.m. of collector area.

4.4.2.4 Biomass

There are varying estimates about biomass utilization in the country. The most recent one is from the 55th round of the National Sample Survey (NSS) in 1999-2000, the estimates derived from this report reveal that in one year (between July 1999—June 2000) around 217 million tonnes (176 + 141) of firewood & wood chips and dung cake was used in the country. As such for 86 per cent of rural households the primary source of cooking energy was firewood and woodchips or dung cakes. In urban areas as well, more than 20 per cent of all households relied mainly on firewood and chips.

Biomass potential could be estimated from three major distinct sources, viz, wood, agro residue and cattle dung. The annual growth of woody biomass in the country has been estimated at 307.4 million tonnes. While forests are estimated to contribute 138 million tonnes, the social forestry, village forests, wastelands and roadside plantations together contribute the rest of 169.4 million tonnes.[23] As wood has multiple uses only a fraction (NSS estimate suggest 176 Mt) of the same is available for energy purposes.

Around 150 million tonnes of surplus agro-biomass resources are generated every year in the country. About half of the surplus residues are burnt in the fields. Therefore, it could be safely presumed that around 70 million tonnes of agro-residues are available.[24] An area of about 107 million hectares has been estimated to be degraded with 64 million hectares categorized as wasteland, which includes degraded forests.[25]

Another bio-resource in India is cattle dung. Country has a livestock potential of around 280 million. With a downside figure of 5 kg dung per cattle per day this cattle population conservatively produces around 500 million tonnes of wet dung annually. With broad estimate of 1 cubic metre biogas per 25 kg wet dung, this dung is sufficient to produce around 20 billion cubic meters (bm^3) biogas per year. Theoretically it would be possible to set up 27 million biogas plants of 2 cubic meter capacity each, implying the dung is sufficient to meet cooking energy needs of the same number of households. However, due to practical limitations such as availability of water, cattle holding per household, etc., Ministry of New and Renewable Energy, Government of India has estimated biogas plant potential at 12 million plants. In addition, Biogas could also be generated from other organic resources such as sewage, municipal solid waste, distilleries etc.

From the above, it could be inferred that the existing utilization of biomass resources in the country is around 387 Mt per year. If 60 million hectare of waste land is developed for energy purposes the biomass availability would be around 687 Mt per year, when biomass growth from conventional routes is considered static.[26]

Table 4.7 Biomass Potential in India

Source	*Existing*	*Useful energy in existing systems*	*Useful energy in alternative systems*
Fire wood & chips (NSS estimate)	176 Mt	7.92 Mtoe (cook stove)	29.04 Mtoe (gasification—thermal application)
Crop residue production (IISc estimate)	70 Mt	2.1 Mtoe (cook stove)	11.55 Mtoe (gasification thermal application)
Cattle dung Cakes (NSS estimate)	141 Mt	2.82 Mtoe (cook stove)	4.45 Mtoe (biogas-cooking)
Total	387 Mt	12.84 Mtoe	45.04 Mtoe
Additional potential if 60 million hectares of wasteland is used for biomass production			
Option 1: Biomass only @ 10 t/ha/year biomass	600 Mt		43.87 Mtoe (electricity—at 20 % system efficiency)
Option 2: Biomass & biofuels –50 % land @ 10 t/ha/year biomass	300 Mt		21.94 Mtoe (electricity—at 20% system efficiency)
–50 % land @ 1.25 t/ha/year biofuels	37.5Mt		8.06 Mtoe (at 20% system efficiency)

4.4.2.5 Other Renewable Energy Sources

4.4.2.5.1 Geothermal Energy

The Geological Survey of India has identified nearly 400 thermal springs distributed in seven geothermal provinces in India. These provinces include The Himalayas: Sohana: West coast; Cambay: Son-Narmada-Tapti (SONATA): Godavari and Mahanadi. Technology Information and Forecasting Council, under Department of Science and Technology, Government of India has published a status report on Geothermal energy utilisation in India.[27] The National Geophysical Research Institute (NGRI), Hyderabad has initiated action to further concretize geothermal energy potential in the country which at present does not appear significant. However, in the long run geothermal energy could become a comparatively viable option for meeting local energy needs in and around the potential locations.

4.4.2.5.2 Ocean Energy

There are three ocean energy technologies, viz, ocean thermal energy, wave energy and tidal energy and India has a coast line of around 6000 kilometre. Ocean thermal energy conversion (OTEC) is a way to generate electricity using the temperature difference of seawater at different depths. The method involves pumping cold water from the ocean depths (as deep as 1 km) to the surface and extracting energy from the flow of heat between the cold water and warm surface water. OTEC Cell at Indian Institute of Technology, Chennai has worked towards developing a feasibility study of setting up a 5 to 8 MW pilot R&D OTEC plant off the main Anandman Islands. As such at present OTEC is a futuristic technology.

Ocean Engineering Centre, Indian Institute of Technology, Chennai, estimated that the annual wave energy potential along the Indian coast is between 5 MW to 15 MW per kilometer. Hence theoretical potential for a coast line of nearly 6000 Km works out to 60,000 MW approximately. However, the realistic and economical potential is likely to be considerably less. Again, wave energy in its present stage of development is a futuristic technology.

Energy can be extracted from tides by creating a reservoir or basin behind a barrage and then passing tidal waters through turbines in the barrage to generate electricity. Tidal energy is extremely site specific and potential sites for tidal power development have already been located. The most attractive locations are the Gulf of Cambay and the Culf of Kachchh on the west coast where the maximum tidal range is 11 m and 8 m with average tidal range of 6.77 m and 5.23 m respectively. The Ganges Delta in the Sunderbans in West Bengal also has good potential at the few locations for small scale tidal power development. The maximum tidal range in Sunderbans is approximately 5 m with an average tidal range of 2.97 m. The identified economic tidal power potential in India is of the order of 8000-9000 MW with about 7000 MW in the Gulf of Cambay about 1200 MW in the Gulf of Kachchh and less than 100 MW in Sundarbans.

Ocean energy technologies are in their initial stage of development and it is unlikely that these sources would cross research and demonstration stage in the medium term upto 2030 or so.

4.4.3 Renewable Energy Policy

At the operational level renewable energy policies could be summed up under four main categories of renewable power, renewable energy for urban and industrial application, renewable for rural areas and new technologies.

4.4.3.1 General Policy Framework

The Ministry of New and Renewable Energy (MNRE) has been providing various support for development of renewable energy technologies in the country. These could be summed up in three broad categories of Fiscal incentives at the central level, specific provisions under Electricity Act, 2003, and financial incentives (subsidy, low interest loan etc.).

4.4.3.1.1 Central Fiscal Incentives

The fiscal incentives include direct taxes—80 per cent depreciation in the first year of the installation of the project, exemption or reduction in excise duty, exemption from central sales tax, and customs duty concessions on the import of material, components and equipment used in renewable energy projects and devices.[28]

The renewable energy devices and systems which have been exempted from excise duty are flat plate solar collectors, black continuously-plated solar selective coating sheets (in cut lengths or in coils and fins and tubes), concentrating and pipe type solar collectors, solar cookers, solar water heaters and systems, solar air heating systems, solar low pressure steam systems, solar stills and desalination systems, solar pumps based on solar thermal and SPV conversion, solar power generating systems, SPV modules and panels for water pumping and other applications, solar crop driers and systems, wind operated electricity generators, their components and parts thereof, water pumping windmills, and aero-generators and battery chargers, biogas plants and biogas engines, agricultural, forestry, agro-industrial, industrial, municipal and urban waste conversion devices producing energy, equipment for utilising ocean waves energy, solar lantern, ocean thermal energy conversion systems, parts consumed within the factory of production of such parts for the manufacture of goods specified above, Solar PV cells.

Income tax benefits are also available for setting up renewable energy projects. Section 80 IA of the Income Tax Act provides benefit to industrial undertakings to set up in any part of India for the generation or generation and distribution of power, including renewable energy.

4.4.3.1.2 Electricity Act, 2003

The most important legislative development which had induced the recent growth in renewable power is the "Electricity Act, 2003" notified in June 2003. It recognizes the role of renewable energy technologies for supplying power to the utility grid as well as in stand-alone systems. Some of the important provisions in the Act in this regard, are as under:

- *Section 3 (1)*: The Central Government shall from time to time, prepare the National Electricity Policy and tariff policy, in consultation with the state governments and the Authority for development of the power system based on optimal utilization of resources such as coal, natural gas, nuclear substances or materials, hydro and renewable sources of energy.
- *Section 4*: The Central Government shall, after consultation with State Governments, prepare and notify

a national policy, permitting stand alone systems (including those based on renewable sources of energy and other non-conventional sources of energy) for rural areas.

- *Section 61 (h)*: The Appropriate Commission shall, subject to the provisions of this Act, specify the terms and conditions for the determination of tariff, and in doing so, shall be guided by the promotion of co-generation and generation of electricity from renewable sources of energy.

  (Majority of the states in India (18 out of 28) have already established the state electricity regulatory commissions (SERC). SERCs are now crucial players in the context of state level policies for renewable.)
- *Section 86 (1)(e)*: One of the functions of the state regulatory commission is to promote co-generation and generation of electricity through renewable sources of energy by providing suitable measures for connectivity with the grid and sale of electricity to any persons, and also specify, for purchase of electricity from such sources, a percentage of the total consumption of electricity in the area of a distribution licensee.

  [The section 86(1)(e) makes it mandatory for the distribution companies to buy certain percentage of the total energy consumption from renewable sources of energy. The SERCs have been given responsibility of determining this percentage or a quota for renewable power.]

4.4.3.1.3 National Electricity Policy

The National Electricity Policy, notified on 12 February, 2005 addresses renewable energy[29] in its Section 5.1.2 (d). It states that reliable rural electrification system will aim at creating:

(a) Rural Electrification Distribution Backbone (REDB) with at least one 33/11 kv (or 66/11 kv) substation in every Block and more if required as per load, networked and connected appropriately to the state transmission system

(b) Emanating from REDB would be supply feeders and one distribution transformer at least in every village settlement.

(c) Household Electrification from distribution transformer to connect every household on demand.

It further states that "Wherever above is not feasible (it is neither cost effective nor the optimal solution to provide grid connectivity) decentralized distributed generation facilities together with local distribution network would be provided so that every household gets access to electricity. This would be done either through conventional or non-conventional methods of electricity generation whichever is more suitable and economical. Non-conventional sources of energy could be utilized even where grid connectivity exists provided it is found to be cost effective."

Section 5.2.20 states that the "Feasible potential of non-conventional energy resources, mainly small hydro, wind and bio-mass would also need to be exploited fully to create additional power generation capacity. With a view to increase the overall share of non-conventional energy sources in the electricity mix, efforts will be made to encourage private sector participation through suitable promotional measures."

Section 5.6.1 of the NEP states that "Special efforts would be made for research, development demonstration and commercialization of non-conventional energy systems. Such systems would need to meet international standards, specifications and performance parameters."

Section 5.12 deals specifically with cogeneration and non-conventional energy sources. The 5.12.1 state that "Non-conventional sources of energy being the most environment friendly, there is an urgent need to promote generation of electricity based on such sources of energy. For this purpose, efforts need to be made to reduce the capital cost of projects based on non-conventional and renewable sources of energy. Cost of energy can also be reduced by promoting competition within such projects. At the same time, adequate promotional measures would also have to be taken for development of technologies and a sustained growth of these sources."

5.12.2 The Electricity Act, 2003 provides that co-generation and generation of electricity from non-conventional sources would be promoted by the SERCs by providing suitable measures for connectivity with grid and sale of electricity to any person and also by specifying, for purchase of electricity from such sources, a percentage of the total consumption of electricity in the area of a distribution licensee. Such percentage for purchase of power from non-conventional sources should be made applicable for the tariffs to be determined by the SERCs at the earliest. Progressively the share of electricity from non-conventional sources would need to be increased as prescribed by State Electricity Regulatory Commissions. Such purchase by distribution companies shall be through competitive bidding process. Considering the fact that it will take some time before non-conventional technologies compete, in terms of cost, with conventional sources, the Commission may determine an appropriate differential in prices to promote these technologies.

5.12.3 Industries in which both process heat and electricity are needed are well suited for cogeneration of electricity. A significant potential for cogeneration exists in the country, particularly in the sugar industry. SERCs may promote arrangements between the co-generator and the concerned distribution licensee for purchase of surplus power from such plants. Cogeneration system also needs to be encouraged in the overall interest of energy efficiency and also grid stability.

4.4.3.1.4 National Tariff Policy

As per the provisions [section 3(1)] of the Electricity Act, 2003, the Ministry of Power, Government of India also notified the National Tariff Policy on January 6, 2006.[30]

Table 4.8 Status of Renewable Energy Quota Across States

S. No.	*State*	*Quota/Renewable Purchase Obligation*	*Time Period*
1.	Andhra Pradesh	Minimum of 5% of consumption of energy	2005-2006 and 2007-2008
2.	Gujarat	2006-07: 1% 2007-08: 1% 2008-09: 2%	2006-2009
3.	Haryana	2006-07: 2% 2007-08: 2% 2008-09: 3% 2009-10:10%	2006-2010
4.	Karnataka	Minimum quantum of 5% and a maximum quantum of 10% of total consumption in a year	2004-05
5.	Madhya Pradesh	Minimum 0.5% of total consumption including third party sales, from wind energy.	2004-07
6.	Uttar Pradesh	5% of total power consumption	—
7.	Maharashtra	2006-07: 3% 2007-08: 4% 2008-09: 5% 2009-10: 6%	2006-2010
8.	Tamil Nadu	10%	2006-09
9.	Rajasthan	Wind— 2%, 2006-07 (7.5%, 2011-12); Biomass—0.5%, 2006-07 (2%, 2011-12)	
10.	West Bengal	2006-07: 1.9% (WBSEB), 1.02% (CESC); 2007-08: 3.8% (WBSEB), 2.03%	
11.	Kerala	5% (2% for Wind)	3 years from 2006
12.	Orissa	3%	2007-08
		Increase @ 0.5% per annum each subsequent year to reach a level of 5% by 2011-12	

Some of the important provisions with regard to non-conventional energy generation are highlighted below:

Section 6.4: Pursuant to provisions of section 86(1)(e) of the Act, the Appropriate Commission shall fix a minimum percentage for purchase of energy from such sources taking into account availability of such resources in the region and its impact on retail tariffs. Such percentage for purchase of energy should be made applicable for the tariffs to be determined by the SERCs latest by April 1, 2006. Such procurement by Distribution Licensees for future requirements shall be done, as far as possible, through competitive bidding process under Section 63 of the Act within suppliers offering energy from same type of non-conventional sources. In the long-term, these technologies would need to compete with other sources in terms of full costs.

Implementation of Section 86 (1) (e) of the Electricity Act, 2003 and Section 6.4 (1) of the National Tariff Policy are underway. Different states are in the process of issuing tariff orders for renewable energy based electricity generation and specifying quota/share for power from renewable energy in accordance with the provisions of the "Electricity Act, 2003".

Many of the states have already issued orders for specifying quotas for renewable energy procurement. These have been summarized in Table 4.8.[31]

4.4.3.1.5 Rural Electrification Policy

In compliance with Sections 4&5 of the Electricity Act, 2003, the Ministry of Power, Government of India has notified the Rural Electrification Policy on 23 August 2006.[32] It views rural electrification as the key for accelerating rural development and states that provision of electricity is essential to cater for requirements of agriculture and other important activities including small and medium industries, khadi and village industries, cold chains, health care, education and information technology. The policy aims at (a) provision of access to electricity to all households by year 2009; (b) quality and reliable power supply at reasonable rates; and (c) minimum lifeline consumption of 1 unit per household per day as a merit good by year 2012.

The approach envisages:

- Grid connectivity through distribution network upto 33/11 or 66/11 KV level with appropriate development and augmentation of sub-transmission and transmission system at higher voltage levels will also be necessary.
- For villages[33] /habitations, where grid connectivity would not be feasible or not cost effective, off-grid solutions based on stand-alone systems may be taken up for supply of electricity so that every household gets access to electricity. Where neither standalone systems nor grid connectivity is feasible and if only alternative is to use isolated lighting technologies like solar photovoltaic, these may be adopted. However such remote villages may not be designated as electrified till the time appropriate solutions are found to provide electricity in these villages to meet the requirements of the definition of village electrification.
- Decentralised distributed generation facilities together with local distribution network may be based either on conventional or non-conventional methods of electricity generation whichever is more suitable and economical. Non-conventional sources of energy could be utilized even where grid connectivity exists provided it is found to be cost effective.

It further states that the State Governments should, within 6 months prepare and notify a Rural Electrification Plan to achieve the goal of providing access to all households. The Rural Electrification Plan should map and detail the electrification delivery mechanisms (grid or stand alone) considering inter alia the available technologies, environmental norms, fuel availability, number unelectrified households, distance from the existing grid etc. and the Plan should also be intimated to the Appropriate Commission.

The electrification services will also cater to the requirement of agriculture and other activities including irrigation pumpsets, small and medium industries, khadi and village industries, cold chains, healthcare and education and IT in a way it facilitates overall rural development, employment generation and poverty alleviation.

Electrification of un-electrified Below Poverty Line (BPL) households would be financed with 100 per cent capital subsidy as per norms of Kutir Jyoti Programme in all rural habitations. Households above poverty line would be paying for their connections at prescribed connection charges and no subsidy would be available for this purpose.

The policy reiterates the definition of an Electrified Village[34] that a village would be classified as electrified based on a certificate issued by the Gram Panchayat, certifying that:

(a) Basic infrastructure such as Distribution Transformer and Distribution Lines are provided in the inhabited locality as well as a minimum of one Dalit Basti/hamlet where it exists;
(b) Electricity is provided to public places like Schools, Panchayat Office, Health Centers, Dispensaries, Community Centers etc.; and
(c) The number of households electrified are at least 10 per cent of the total number of households in the village (Though the requirement for categorizing a village as electrified is electrification of atleast 10 per cent households, this policy aims at providing access to all the households).

In order to manage local distribution, deployment of franchisees for management of local distribution in rural areas is considered necessary in order to ensure revenue sustainability and improve services to the consumers. Franchisees for the management of rural distribution could be non-governmental organisations (NGOs), users' associations, cooperatives or individual entreprenures. A franchisee would be responsible for distribution of electricity within an identified contiguous area for a prescribed duration and for collecting revenues directly from the consumers.

4.4.3.1.6 Foreign Direct Investment

Government of India is encouraging foreign investors to set up such power projects on Build-own and Operate basis. For entire renewable energy sector, 100 per cent foreign direct investment is allowed under the automatic route without prior information to the government. No prior approval of government is required to set up an industrial undertaking with FDI/NRI/OCB investment. The investor can bring funds directly, incorporate Indian company, allot shares to foreign investors and inform RBI within 30 days.

4.4.3.1.7 Financial Incentives

Direct financial incentive is provided for application and adoption of renewable energy technologies in country. The level on quantum of support generally varies from year to year, depending upon availability of financial resources for the particular year in government budget, level of technology maturity, socio-economic condition of the user group, etc.

4.4.4 Renewable Energy Programmes

4.4.4.1 Wind Power

The Indian wind power programme is private sector driven, supported with fiscal incentives and preferential tariff. As of March 2007, 7092 MW of wind power installed capacity was set up in the country.

4.4.4.1.1 Technology

The Indian wind turbines are globally competitive. Initial efforts in late 1980s were to develop wind turbines indigenously and in this process 55 KW wind turbine was developed by Bharat Heavy Electricals Limited (BHEL). Subsequently, with focus on private sector participation and fast developments in other parts of the world, the focus shifted to acquiring technologies through technology transfer or joint venture route. In this process an overall indigenization of about 80 per cent have been achieved. As wind energy programme is mainly market driven, supported by conducive policy framework, the technology development efforts are primarily done by industry within India and through its subsidiaries and/or joint venture collaborators in other countries. However, Department of Scientific and Industrial Research (DSIR) has been making efforts to develop indigenous wind turbines of 500 KW capacity.

Table 4.9 Major Wind Turbines in Indian Market in 2006

Sl. No.	*Manufacturer*	*Model*	*Capacity (KW)*
1.	Enercon India Limited	E-33	330
		E-48	800
2.	GE India Industrial Pvt. Ltd	1.5s	1500
3.	NEG Micon (India) Pvt. Ltd	NM48	750
		NM82/V82	1650
		NM72C	1500
4.	NEPC India Ltd	NEC 225	225
5.	Poineer Asia Wind Turbines	Gamesa Eolica	850
6.	Suzlon Energy Ltd	N3330	300
		N3335	350
		S64	1000
		S64	1250
		S66	1250
7.	Vestas RRB India Ltd.	V39	500

The wind turbines offered in Indian market are of two basic types the geared wind turbines with induction generator and gearless wind turbines with synchronous generator. In 2006, there were around 7 major companies involved in the manufacturing/supplying wind turbines in India as given in Table 4.9.

4.4.4.1.2 C-WET

The MNRE has set up the Centre for Wind Energy Technology (C-WET), Chennai, Tamil Nadu in 1998 as an autonomous institution. Its main activities include wind resource assessment, testing, development of standards and certification, research and development and human resource development. As a part of wind resource assessment it has been collecting wind data from different states and working towards further refining wind data of the country. It has undertaken research projects for design and development of Intelligent Controller for wind turbines and design and maintenance of gearboxes of wind turbine. Its current activities include grid interconnection issue, design methodologies for Wind Turbine Blades and development of wind turbines for low and moderate wind regimes.[35]

4.4.4.1.3 Cost

The installed cost per MW of wind power projects vary from location to location as well as from supplier to supplier. This sector is highly competitive and hence the actual costs are difficult to get. A review of tariff orders, specifying the rate at which the utilities would buy power generated from wind power projects, issued by different state electricity regulatory commissions show a diversity in capital costs varying from Rs. 40—50 million/MW. Based on these costs and the capacity utilisation factors the regulatory commissions have estimated buy back tariffs which will ensure about 16 per cent return on equity. Tariff offered in different states is given in Table 4.10.

Table 4.10 Tariffs Offered to Wind Power Projects in Different States

State	*Rate at which utilities would buy the wind power*
Andhra Pradesh	Rs. 3.37/kWh with 5 per cent annual escalation in rate with base year as 2004
Haryana	Rs. 4.08/kWh for the year 2007-08 with 1.5% annual escalation
Madhya Pradesh	Rs. 3.97 (with decrease of 7 paisa upto 4th year) then fixed at Rs.3.30/kWh from 5th year onwards uniformly for 20 years (w.e.f.11.6.2004)
Karnataka	Rs. 3.40/kWh for the first year with no escalation
Uttar Pradesh	Rs. 2.50/kWh for the year 05-06 with 4 per cent annual escalation
Maharashtra	Rs. 3.50/kWh for the first year with increment of 15 paisa every year for thirteen years
Rajasthan	Rs. 3.25/kWh (base year 05-06) with 6 paise annual escalation for 9 years.(upto 2014-15 then fixed at Rs 3.79/kWh)
Tamil Nadu	Rs. 2.75/kWh for old plants Rs. 2.90/kWh for new plants
Gujarat	Rs. 3.37/kWh fixed for 20 yrs.
Kerala	Rs. 3.14/kWh fixed for 20 years
West Bengal	To be decided on case to case with a cap of Rs. 4/kWh

4.4.4.2 Small Hydro Power

In Indian context, hydro power projects upto station capacity of 25 MW have been termed as Small Hydro Power (SHP). As of March 2007 a total of 1975 MW of total small hydro power had been installed in the country. Many states have brought in policies, regulations and tariff orders to further give push to small hydro development.

During initial years utpo 1992-93 demonstration projects were set up in various states to regenerate interest of State Governments along with provision of financial incentives like capital subsidy of up to 50 per cent of the cost of project cost. Policy initiatives for private participation in SHP started during 1993-94, keeping in view the overall policy of Government of India to encourage private sector participation in the field of power generation. In addition to MNRE's fiscal and financial incentive, the UNDP (United National Development Programme)/ GEF (Global Environment Facility) assisted hilly hydro project operationalized in 1994 through MNRE for setting up of demonstration projects and preparing a national plan for the development of small hydel projects in the entire Himalayan region laid foundation for taking up larger project portfolio.

4.4.4.2.1 Technology

India has a well-established manufacturing base for the full range and type of small hydro equipment. SHP project comprises civil works and electro-mechanical equipment. The civil works includes a simple construction for regulated discharge to the feeder channel, an open tank designed to maintain a balance in the power output by providing a steady design head for the project, a penstock for steady and laminar water flow into the turbine, and the electro-mechanical works which include a turbine typically a Pelton or Turgo design for high-head applications and a Kaplan or Francis design for low head applications, a generator, usually a synchronous design for larger SHP sites and self excited induction design for low-power SHP applications. In addition, electronic load controller, depending on whether the turbine and generator operate on full or varying load conditions is required. SHP technology is well established in the country and over the years focus is on standardization, automation and efficiency improvement. There are around 11 major SHP component manufacturers in the country, besides many NGO and individuals group have expertise in assembling SHP from the first principle.

4.4.4.2.2 Cost

The installed cost per MW of small hydro power projects are very site specific. Tariff orders across states implementing small hydro projects show that the capital cost varies in the range of 36-55 million Rs/MW. Based on these capital cost benchmarking buy back rates have been offered by the states. Table 4.11 gives the tariffs offered to small hydro power projects in different states.

Table 4.11 Generation Tariffs Offered for Small Hydro Projects

State	*Rate at which utilities would buy the small hydro power*
Andhra Pradesh	Rs. 2.60/kWh for the first year Royalty charges will be paid by APTRANSCO and Discom's to GoAP. Re. 0.39/kWh up to 5 years. Re. 0.78/kWh beyond 5 years and up to 10 years. Rs. 1.17/kWh beyond 10 years.
Haryana (upto 2MW)	Rs 3.67/kWh for 2007-08 with 1.5% annual escalation
Karnataka	Rs. 2.80/kWh for the first year with no escalation No Royalty charges.
Uttar Pradesh	Rs. 3.39/kWh Royalty charges will be paid by distribution licensees to GoUP
Uttarakhand	For projects less than 1 MW and for less than 25 MW it will do according to regulation and on case to case basis. For projects greater than 1 MW the guiding regulation will be "tariff will be determined as the weighted average of power allocated to the state from central government stations. All other related provision remains unchanged as per the regulation of 2004". If any generating company wants to get the tariff determined in accordance to the regulation of 2004 then he could approach the commission.

(*Contd.*)

State	Rate at which utilities would buy the small hydro power
Maharashtra	Rs. 2.84/kWh in the first year, which increases by Re. 0.03/kWh every year till the debt repayment, is over i.e. in the 10th year. The tariff shall remain constant at Rs. 3.11/kWh between the 10th year and the 15th year after which it again increases annually at a constant rate of Re. 0.03/kWh. Royalty charges are passed through.
Himachal Pradesh	Rs. 2.50/kWh with the base year 2001. Provided that the tariff shall be indexed at 50% of the annual inflation rate of the Consumer Price Index. The inflation upto 50% shall be absorbed by the generator of the renewable energy and 50% inflation shall be added to the base rate (as given in the 'Draft Regulations for power procurement from renewable sources, November, 2005').

4.4.4.3 Solar Photovoltaics

Solar PV sector programmes in India consists of: (a) subsidy driven market, (b) niche product market, and (c) open market. The Government of India started deploying solar PV systems as part of rural development programme since 1970s. Initially, the SPV systems were deployed under 100 per cent capital subsidy which was reduced subsequently. The present focus of the government supported dissemination programmes is on demonstration and utilization of lighting systems, stand alone power plants and other specialized systems in the country, meeting electricity needs of such remote villages where other renewable options are available, water pumping applications and applications in the urban Areas. The systems which are being stalled under this program include, solar lanterns, solar home systems, small solar photovoltaic power plants and solar water pumping systems. In addition there exists non-subsidy market for niche applications like railway signalling, power packs for rural telecom equipments, oil platforms and cathodic protection of oil pipe lines. Indian manufacturers have been exporting solar PV modules and systems to nearby countries. Exports to US and Europe are on rise in recent years. By the year 2007-2008 the total solar cell manufacturing capacity in the country was 163.5 MWp and the module capacity was 185 MWp. However, of an aggregate capacity of 320 MWp solar PV modules produced in the country by March 2007, over 70 per cent were exported. The domestic market is limited to only around 10 MWp per year. There are 9 manufacturers of solar cells, 17 of PV modules and another 80 of PV systems.

4.4.4.3.1 Technology

Crystalline silicon based solar cell and module manufacturing technology is most commonly used technology in India and their field efficiencies are comparable to global average. Although two manufacturers have developed amorphous silicon solar cells these are not commercially available in India. Seven Indian standards for PV modules and other related aspects have been developed. Solar cells manufacturing and module manufacturing are commercialised technologies. In 2006, there were only two MNRE funded research projects for developments of new PV concepts for development of nano and microcrystalline silicon based low cost, high efficiency thin film solar cells and optimization of the performance of multi junction thin film solar cells.

4.4.4.3.2 Economics of the Solar PV

SPV systems have steadily gained power generation market share due to their ability to produce electricity with no moving parts, no fuel requirements, no water requirement and no noise. The environmental impact of SPV technology is nil at the point of use. Modules produce electricity silently and do not emit any harmful gases during operation. Silicon, the basic photovoltaic material used for most common solar cells, is environmentally benign. The modular nature of SPV, which allows systems to be configured to produce power from Watts to Mega-Watts, gives it a unique advantage over other technologies. Theoretically, any amount of electricity generated from solar PV could be supplied to grid. However, in practice the configurations used for grid-interactive systems is in KW to MW range. If the capacity factor is assumed at 20 per cent with 6 hrs/day average power generation at peak level for 300 days and the solar modules are rated at design operating conditions of 25°C ambient temperature and insolation of 1000 W/m^2, the cost of the electricity generated will be Rs. 15.60-17.60/kWh. In the above calculation it has been assumed that solar modules are deployed to face south and are inclined at an angle equal to latitude to achieve maximum solar energy collection throughout the year.

In the present context, solar PV is mainly used for off-grid applications in the country. The costs of PV systems vary widely and depend on system size, location, off-grid requirement and also technical specifications. The cost of systems has two main components, viz solar module and balance of system cost including battery. A comparison between the system cost based on actual field data for solar PV based power generation in KW range along with competing conventional fuel alternative reveal that with around Rs. 25/kWh solar PV based electricity generation is the cheaper option if compared with petrol and LPG based power generation systems in the same range for off-grid applications.

4.4.4.4 Solar Thermal Power

Solar thermal power generation technologies were tried on small demonstration scale in the past but no major development has taken place in this area. There is no commercial solar thermal power plant in operation in India, neither there is any research group that is working in this direction. Over the years solar thermal power generating systems have improved in terms of cost reduction and are seen as a reliable option. At the global level, electricity production has been improved by improving operation and maintenance procedures, thereby reducing levelized electricity cost (LEC). Besides many improvements, two major technological developments are under way: integration of parabolic trough power plants in combined cycle plants and direct steam generation in the collectors' absorber tubes. By using direct solar steam generation, the heat transfer/water heat exchanger will no longer be required. Thus, by reducing investment costs and at the same time increasing system efficiency, a reduction of LEC is expected.

4.4.4.5 Solar Heat

In India the solar thermal systems are used both in the domestic as well as commercial and industrial sector. Solar thermal energy can be utilized for various end uses mainly for solar water heating system, solar energy for industrial application, and solar cooking. It is estimated that approximately 1000 solar domestic water heaters (2000 m^2 of collector area) can contribute to a peak load saving of 1MW. There are now 79 manufacturers of solar flat plate collectors with approval from the BIS (Bureau of Indian Standards) for their products. Solar flat plate collectors with copper absorber having black chrome selective coating are most common in India. Indigenously developed selective coating technology is commonly used. There are about five manufacturers of selective coating technology. Typically these systems are in the range of 100 lit per day to 500 lit per day capacity. Vacuum tube based solar collectors manufactured in China are slowly becoming popular in India.[36]

There are major industries where boilers supply process heat in the form of either steam or hot air upto a temperature of 150°C. These industries include dairy, food processing, textiles, hotels, chemicals, breweries and distilleries. It has been estimated that around 60 per cent of the thermal energy consumed in the industry is used to process end products. Few large scale solar thermal systems for industrial process heating have been installed in the country. These include a 100,000 lit per day capacity solar system installed in a fertiliser unit for supplying pre-heated water for boiler. Besides this many large institutions such as hostels and religious places use hot water in large quantities. Thus industries and institutions are also major market segments for solar thermal systems.

The solar cooker programme started in India with various promotional activities including subsidy. A box solar cooker is a slow cooking device useful for small families. It can cook 4 dishes at a time and save 3-4 LPG cylinders in a year if used regularly. The dish solar cooker is a fast cooking device useful for homes and small establishments. It can cook almost all types of food for about 10-15 people. The cooker can save around 5 to 10 LPG cylinders in a year depending upon its use in homes or small establishments.

The community solar cooker for indoor cooking has a large automatically tracked parabolic reflector standing outside the kitchen; it reflects the sunrays into the kitchen through an opening in its north wall. A secondary reflector further concentrates the rays on to the bottom of the cooking pot, which is painted black. It can cook all types of food for about 40-50 people and can save up to 30 LPG cylinders in a year with optimum use.

The solar steam cooking system comprises automatically tracked parabolic reflectors installed in a series and parallel combination that generates steam for cooking in community kitchens. It can cook food for thousands of people in a short time depending upon its capacity. It is normally installed in conjunction with a boiler that uses conventional fuel when necessary. Solar dish based steam generation system are slowly becoming popular among the institutes where large scale cooking is done using steam as heating media.

4.4.4.5.1 Cost Economics

The useful energy provided by a solar thermal system depends on the availability of solar radiation and ambient conditions. This apart, the actual amount of fuel/electricity saved by the use of solar thermal systems depends upon the efficiency of utilization of the fuel/electricity for heat purposes and calorific value of the fuel. The cost of solar heat systems depends on specific end use requirements and size of the system. The economic viability or the pay back period depends on type of fuel saved, location of solar system which in turn decides its output based on available solar energy, and percentage of utilisation. As far as the solar water heaters are concerned the pay back periods are typically 2 to 3 years and systems are economically viable in areas where there is a continuous requirement of hot water; for example in hotel, hospital etc. when electricity is saved.

4.4.4.6 Biomass-based Power

Indian biomass programme focuses on biomass-based power generation; cogeneration; research and development; and biomass resources assessment. The objectives are to optimally utilise a variety of biomass material such as agricultural residues, agro-industrial residues, forestry-based residues and dedicated energy plantations by adoption of state-of-the-art advanced technologies. The aggregate biomass combustion based power and sugar-cogeneration capacity by March 2007 was around 1140 MW, with 600 MW of cogeneration and the rest biomass power. In the area of small-scale biomass gasification, a total capacity of 76 MW has so far been installed, mainly for stand alone applications.

4.4.4.6.1 Technology

Globally competitive biomass combustion and gasification technologies have been developed in the country. Some of the thrust areas for technology advancements in the country include development of high-pressure boilers and new turbine configurations; cost effective biomass handling, storage and drying systems; feasibility of blending or co-firing of mix of biomass materials, or with fossil fuels; installation and evaluation of Pilot Dendro-thermal Power Plant; and indigenous technology development/adaptation of advanced biomass gasification involving gas turbines in combined cycle mode.

4.4.4.6.2 Cost

The installed cost per MW of biomass projects depends upon the specific technology route applied. Tariff orders across states implementing biomass power and co-generation projects show that the capital cost varies in the range of 35-40 million Rs. MW. Based on these capital cost benchmarking by states' a preferable buy back rates have been offered by the states.

Table 4.12 Tariffs offered to Cogeneration/Biomass Power Projects

State		*Rate at which utilities would buy the cogeneration/biomass power*
Andhra Pradesh	Cogen	Fixed cost Rs. 1.72/kWh in 1st year reducing to 0.90 in 10th year + variable cost Rs. 1.02 in 2005-06 escalating to Rs. 1.24 in 2008-09
	Biomass power	Fixed cost Rs. 1.61/kWh in 1st year decreasing to 0.87 in 10th year. The variable cost for 2004-05 was Rs. 1.27 escalating to Rs. 1.54 in 2008-09.
Haryana	Cogen	Rs. 3.74/kWh with 2 per cent escalation
	Biomass power	Rs. 4.0/kWh with 2 per cent escalation
Maharastra	Cogen	Rs. 3.05/kWh with 2 per cent escalation
	Biomass power	Rs. 3.04/kWh in first year escalating to Rs. 3.34 in 10th year
Karnataka	Cogen	Rs. 2.87/kWh in 1st year with escalation reaching Rs. 3.06/kWh in 10th year
	Biomass power	Rs. 2.93/kWh in first year escalating to Rs. 3.10 in 10th year
Tamil Nadu	Cogen	Rs. 3.15/kWh
	Biomass power	Rs. 3.15/kWh

4.4.4.7 Biogas

India's quest for biogas technology dates back to 1897, when the Matunga Leper Home in Mumbai utilized human waste to generate gas for meeting its cooking and lighting needs. In the decades to follow, biogas research and development occurred at various levels. At one level, the Khadi and Village Industries Commission (KVIC), a government-sponsored umbrella organization for rural industrialisation, was engaged in developing new efficient models for biogas generation from biodegradable waste. At another level, researchers at the Indian Agricultural Research Institute (IARI) in Delhi were developing efficient systems for anaerobic digestion of cattle dung. Several models were developed by 1950, the most notable being Grama Laxmi III, which became the prototype of the KVIC floating dome model. With this successful design, KVIC began field implementation in 1961 and installed over 6,000 biogas plants around the country by 1973. Mrs. Gandhi, the then Prime Minister of India even made a reference to biogas in meeting rural energy needs in her Independence Day address to the nation in 1974.

4.4.4.7.1 National Project on Biogas Development

Later in 1981 a National Project on Biogas Development (NPBD) was launched by the Ministry of Agriculture which was transferred to the newly created Department of Non-conventional Energy Sources, subsequently. NPBD initially focused on 100 selected districts, and a modest target of 400,000 biogas plants was set for the Sixth Five-Year Plan period. National Project on Biogas Development (NPBD) has been one of the well organized and systematic programmes that has been under implementation since early 1980's and by March 2007, around 3.9 million family type biogas plants have been set up in the country. It is the only programme that has put cooking in rural areas on technological ladder and has made cooking a pleasure with associated social and environmental benefits including zero indoor pollution. Although there are various claims about success, the national census in 2001 has revealed that around 1 million households are meeting their cooking energy needs mainly through biogas. There are other reports, which say that for around 55 per cent of plant owner biogas is predominant fuel. As such it may be inferred that Indian biogas programme is one the most successful programmes if we compare with other such programmes world wide and also with other programmes implemented in rural India. The ultimate goal of Indian biogas programme is to set up biogas plants in around 12 million rural Indian households that have enough cattle to maintain a regular supply of dung. Depending on family size, this would mean a regular supply of fuel for 60 to 85 million people.

One of the major barriers for large-scale dissemination of family type biogas plants is upfront cost of installation of biogas plants. Central government financial assistance has been reducing over the years. Apart from this the annual target for biogas has also been declining year after year, as in 2005-06 the initial national target was only 25,000 plants. In many places there is demand for biogas plants, but financial assistance is not available due to limited target (e.g in Bagepalli CDM-PDD it has been observed that the central and state government supported only 500 biogas plants in Kolar district in 2005, whereas the demand may easily exceed 50,000 plants).[37]

NPBD in its present form does not fulfil the aspirations and expectations as initially envisaged. In the absence of an internal mechanism of repairing and overhauling, the problems have compounded over the years. In the haste to spread the technology to every nook and corner of the country, proper assessment of socio-economic and ecological factors have not been conducted, which has led to dilution and inefficient utilisation of resources. Feedback on operational problems related to running of biogas plants was not adequately addressed with the result that the fixed-dome Deenbandhu model became the lone surviving design, although it has certain structural and social limitations which makes it unfit for propagation in certain regions of the country. Users did not have any design options to choose from.

Research on alternate feedstocks did not go beyond the laboratory stage.

Biogas is mainly a programme that serves twin objectives of meeting cooking energy needs of the rural areas and in producing enriched manure. Conventional fuel based cooking energy systems such as LPG and Kerosene are subsidized in the country. Over the years average subsidy on biogas plants have been reduced progressively and also the number of such plants supported on yearly basis are reducing. It may be mentioned that whereas there is one time subsidy on biogas plants, the subsidy on LPG and Kerosene is on regular basis. As per the 55th round of the National Sample Survey (NSS), LPG and Kerosene penetration is very poor in rural areas and only 5 per cent and 2.7 per cent households uses LPG and Kerosene respectively for cooking purposes. Thus there is wide disparity in providing support for cooking energy services in rural and urban areas. A rough estimate indicates that the subsidy provided on LPG to the higher income group households in Delhi is much higher than total quantum of subsidy provided for biogas based cooking in the rural India.

4.4.4.8 Bio Fuels

The three bio-fuel options being considered in the country are straight vegetable oils (SVOs), bio-diesel and bio-ethanol. India's bio-fuel programme is primarily based on bio-ethanol derived from sugarcane molasses and bio-diesel derived from non-edible oil seeds.

4.4.4.8.1 Bio-Ethanol

India is the second largest producer of ethanol in Asia. Installed ethanol production capacity amounts to approximately 2.7 billion litres but capacity utilisation rates are usually rather low (total ethanol production in 2004 amounted to 1.7 billion litres). In 2004, around 100 million litres of ethanol were used for blending with petrol. Assuming that the ethanol programme will be implemented in those States and Union Territories originally envisaged by the government, the projected demand for fuel ethanol will be 1.5 billion litres in 2010 if E-10 is sold.

Ethanol production in the country is constrained because of its dependence on a single source—sugarcane molasses. During the period March 2003 to September 2004, 0.37 million kl of fuel ethanol was purchased by the oil industry as a part of the 5 per cent ethanol blending programme. In 2005, the fuel ethanol prices were around Rs. 19.50 per litre.

In the year 2000 the Ministry of Petroleum and Natural Gas took up pilot projects at three locations in the major sugar-producing states of Maharashtra and Uttar Pradesh to study the related aspects of blending ethanol with petrol and its use. Ethanol-blended petrol, supplied through more than 300 retail outlets under these pilot projects, proved successful. Subsequently, Government of India issued a notification on 12 September, 2002 mandating supply of 5 per cent ethanol-blended petrol in nine States[38] and four Union Territories[39] with effect from 1 January 2003. Following a drought, the 2003/04 and 2004-05 sugar crop and therefore also molasses outputs was unusually low, which resulted in sharply increased feedstock prices for ethanol production. It became more lucrative for local producers in India's Southern States to concentrate on industrial and potable alcohol. The ethanol blending obligation was temporarily suspended in the autumn of 2004.

The Policy measures currently in place include an excise tax reduction for E-5, the obligation to blend all petrol with 5 per cent ethanol in certain regions since January 2003 and government regulation of the ethanol selling price on the basis of ethanol production costs. In 2006, 5 per cent ethanol blends were used in 10 sugar producing states and 3 contiguous Union Territories. In addition to the federal moves, several Indian States have also attempted to support local ethanol production through the use of additional fiscal measures. In the next phase, supply of 5 per cent ethanol-blended petrol is proposed to be extended to the whole country. Subsequently the percentage of ethanol mixture in petrol is planned to be increased to 10 per cent.

4.4.4.8.2 Bio-Diesel

India has more than 300 different species of trees, which produce oil-bearing seeds. Around 75 plant species, which have 30 per cent or more fixed oil in their seeds/kernel, have been identified and listed. The promising non-edible sources are: Jatropha curcas (Ratan Jyot), Pongamia pinnata (Karanja), Melia azadirachta (Neem), and Shorea robusta (Sal). Traditionally, collection and selling of tree based oilseeds (TBOs) were generally carried out by poor people for use as fuel for lighting. There exist small traditional markets for trading of tree borne oil seeds and oils. The present production of oil from TBO is estimated at 60,000 tonnes/annum and these oils are used in soaps, varnishes, lubricants, candles, cosmetics, etc. In addition, these oils are being used to conducting trials on vehicles and lab experiments. In 2006 the market price of bio-diesel varied from Rs. 55—110 per litre. The cost of jatropha bio-diesel from jatropha was particularly high i.e. Rs. 80-110 as jatropha seeds were in high demand for raising new plantations.

Table 4.13 Selected State-wise Jatropha Nursery Programme in 2006-2007[41]

States	*Number of Seedlings (in million)*
Gujarat	15
Andhra Pradesh	35.833
Chhattisgarh	62.5
Rajasthan	7.5
Himachal Pradesh	5.0
Tamil Nadu	35.0
Sikkim	5.0
Assam	5.0
Tripura	10.0
Total	180.833

The National Oilseeds and Vegetable Oils Development (NOVOD) Board, Ministry of Agriculture projected that by the year 2008-09, Jatropha plantations would extend over an area of about 3.1 million hectares. These plantations would produce an estimated 2.914 million tonnes bio-diesel (at the rate of 0.94 tonnes per hectare) annually. This level of bio-diesel production in the country would result in a likely saving of foreign exchange equivalent to about Rs. 95,000 million per annum at the current import parity price of diesel in international market of Rs. 27.87/ litre of HSD.[40]

Table 4.14 Some Jatropha Initiatives

Name of the Agency	*Plantation details*	*Remarks*
National Oilseeds & Vegetable Oils Development Board (NOVOD)	7,500 ha	Model plantations in 21 states
BP-TERI	8,000 ha	Year of plantation: 2006 onwards
Uttaranchal Bio-fuel Board	10,000 ha	Year of plantation: 2005
Chhattisgarh	80 million seedlings	Year of plantation: 2005
CSMCRI-Daimler-Chrysler project	20 ha in Gujarat + 20 ha in Orissa	Year of plantation: 2003, onwards
IOC-Indian Railways	1,10,000 saplings at Surendra Nagar, Gujarat	

As per the Wastelands Atlas of India—2005, brought out by the Department of Land Resources, Ministry of Rural Development in collaboration with the National Remote Sensing Agency (NRSA), Hyderabad using satellite imagery data, an area of 55.269 million hectares in the country is estimated to be wastelands/barren lands. Out of this, 12.117 million hectares is unculturable, being barren, rocky, snow-covered, etc. Remaining 43.152 million hectares is culturable. This includes 30.497 million hectares of non-forest culturable waste land and 12.655 million hectares of degraded forestland. Out of the 43.152 million hectares culturable wastelands, it is estimated that 32.296 million hectares would be suitable for plantation of Jatropha.[42]

4.4.4.8.3 Technologies

The process of bio-diesel manufacture involves pre-treatment of crude vegetable oil followed by the process of trans-esterification using methanol or ethanol. Bio-diesel is separated from glycerol and it is then washed and dried to obtain the final product. Glycerol is a by-product of the process. Technology for the preparation of bio-diesel from low free fatty acid (FFA) content and high quality oils is well known. The challenge in India is to handle multi feedstocks with high FFA (jatropha and pongamia have high FFA compared to several edible oil).

Several Indian R & D institutions are working on small-scale (up to few thousand lpd capacities) trans-esterification technology. CSMCRI, Bhavnagar has developed acid and base catalysis process to produce high quality bio-diesel. Acid catalysis is preferable as a pre-treatment option for vegetable oil having high free fatty acid content. The process can be used for capacities ranging from 100 lpd to 2000 lpd. One batch operation takes 8 hours for completion. The typical cost of 500 lpd capacity plant is estimated at Rs. 6 million, including the civil cost.

Indian Institute of Chemical Technology, Hyderabad, has developed a batch process with the traditional catalysts both for low and high FFA vegetable oils. The technology developed by Indian Oil Corporation (IOC) has been transferred to M/s Venus Ethoxyethers. Under a World Bank Development Marketplace supported project, being implemented in Orissa by CTx Green and Gram Vikas, very small capacity (5-25 lpd) trans-esterification plants, suitable for operation in remote villages are being developed. Apart from these institutions, Delhi College of Engineering, Punjab Agriculture University, Indian Institute of Petroleum are some of the other institutes working on bio-diesel production technology.

4.4.4.9 Hydrogen

Government of India had set up a special task force in July 1976 under the Department of Science and Technology to initiate an action plan for the development and implementation of hydrogen energy program in the country. DST identified hydrogen energy as one of the thrust Areas in Engineering Sciences in 1982 and some research projects were supported to universities, technical institutes and national research laboratories. The Technology Information, Forecasting and Assessment Council (TIFAC), under DST has supported projects to various organizations as a part of Technology Vision 2020 mission which were aimed at transition to the application of clean energy fuels including hydrogen. In parallel, Ministry of New and Renewable Energy also funded research projects on different aspects of hydrogen energy technology development. Under these projects upto 50 KW phosphoric acid fuel cells have been developed and demonstrated. Preliminary research for hydrogen production through biological route has also been initiated. In addition, metal hydrides technology for hydrogen storage developed within the country is globally competitive. Research for carbon nano-structures for hydrogen storage is at initial stage. Solid Oxide Fuel Cells are at early development stage. As such hydrogen energy technology development is at its initial stages in the country.[43]

India is member of the International Partnership on Hydrogen Economy (IPHE) set up in Washington D.C. in November, 2003. In Fourth Steering Committee Meeting of IPHE held in Kyoto in September, 2005 in which India presented its achievements in Hydrogen Energy and Road Map. The National Hydrogen Energy Board functions as an overall body for guiding preparation and implementation of National Hydrogen Energy Roadmap (NHEP) and also for creation of policy framework cutting across different sectors and also guiding research and development programme in Hydrogen and Fuel cells. Two major initiatives have been proposed in NHEP for hydrogen based future transport and power generation.

NOTES

1. Planning Commission 2002: http://planningcommission.nic.in/plans/planrel/fiveyr/10th/volume2/v2_ch7_3.pdf
2. Integrated Energy Policy Report, Planning Commission, 2006
3. Ministry of Coal, 2005.
4. Key World Energy Statistics, IEA, 2006.
5. http://powermin.nic.in/JSP_SERVLETS/internal.jsp as accessed on 8 March, 2007.
6. http://powermin.nic.in/JSP_SERVLETS/internal.jsp as accessed on 8 March, 2007.
7. General Review, 2006, Central Electricity Authority Ministry of Power, Government of India.
8. Central Electricity Authority Ministry of Power, Government of India (2006): figures in parentheses are for 2004-05.
9. General Review, 2006, Central Electricity Authority Ministry of Power, Government of India.
10. Census of India, 2001.
11. Office of Registrar General & Central Electricity Authority as referred in Rajya Sabha Unstarred Question No. 3123, dated 15.05.2006.
12. Planning Commission, 2006.
13. Planning Commission: Plan documents & mid term reviews reports. Expenditure at current price at base year of the respective Plans.
14. Adopted from Integrated Energy Policy Report, Planning Commission, 2006.
15. Adopted from Integrated Energy Policy Report, Planning Commission, 2006.
16. Ministry of Petroleum and Natural Gas, 2004.
17. Planning Commission, 2006.
18. Report of the fuel and power sector, submitted to the National Committee on Science & Technology, Department of Science and Technology, Government of India, 1976.
19. Government of India Cabinet Secretariat Resolution as amended under No 64/1/3/82-Cab. Dated 23 October, 1982.
20. http://planningcommission.nic.in/plans/planrel/app11th_24.pdf
21. http://www.cwet.tn.nic.in/html/departments_wra.html
22. Central Electricity Authority, 2006.
23. As referred in Mission Document: National Mission on Decentralized Biomass Energy for Villages and Industries (January, 2006).

24. Indian Institute of Science, Bangalore, 2005.
25. Wasteland Atlas of India, Department of Land Resources (Ministry of Rural Development).
26. Basic assumptions in this calculation include: Calorific value per kg Wood: 4500 Kcal; agro waste: 3000 kcal; and dung cake: 2000 kcal; Conversion factor: 10 per cent in cook stoves.
27. http://www.tifac.org.in/offer/tlbo/rep/TMS153.htm as accessed on 12 March, 2007.
28. http://www.mnes.nic.in The Gazette of India, Extraordinary, Part-I-Section-1, No 23/40/2004-R&R (Vol.II), Ministry of Power Government of India The Gazette of India, Extraordinary, Part-I-Section-1, No 23/2/2005-R&R (Vol.III), Ministry of Power, Government of India.
31. Websites of respective state electricity regulatory commissions
32. The Gazette of India, Extraordinary, Part-I-Section-1, No 44/26/05-RE(Vol-II), Ministry of Power Government of India.
33. Villages means census villages.
34. Ministry of Power's O.M. No.42/1/2001-D(RE) dated 5th February, 2004.
35. http://www.cwet.tn.nic,in as accessed on 11 March, 2007.
36. TEDDY 2003-04.
37. http://cdm.unfccc.int/Projects/DB/DNV-CUK1131002343.1/view.html
38. Andhra Pradesh, Goa, Gujarat, Haryana, Karnataka, Maharashtra, Punjab, Tamil Nadu and Uttar Pradesh.
39. Chandigarh, Dadra & Nagar Haveli, Daman & Diu and Pondicherry.
40. Lok Sabha starred question no 431 answered on 25.08.2006
41. Lok sabha unstarred question no. 653, dated on 24.11.2006
42. Lok Sabha starred question no 431 answered on 25.08.2006. However, the West land Atlas of India has estimated wasteland at 64 million hectares.
43. Presentation Towards Hydrogen Energy Economy in India by Dr S.K.Chopra, Former Special Secretary, MNRE.

Chapter 5

Renewable Energy and Environment

Significant growth in energy consumption based on fossil fuel resources has made an impact on the earth environment. As such local pollution from energy systems remains a threat to the health of the living environment. Now pollution resulting from combustion of fossil fuels has resulted in a global concern. The planet's climate is changing as the result of a build up of greenhouse gases in the atmosphere. The cost arising due to aforesaid impacts of energy generation, transmission and consumption are generally not accounted in determining cost of energy, and regarded as external cost of the delivered energy.

An external cost is known as an externality, arises when the social or economic activities of one group of persons have an impact on another group and when that impact is not fully accounted, or compensated for, by the first group. In other words externalities are defined as benefits or costs generated as an unintended by-product of an economic activity that do not accrue to the parties involved in the activity and where no compensation takes place. EU defined fuel cycle externalities as "the costs imposed on society and the environment that are not accounted for by the producers and consumers of energy, *i.e.* that are not included in the market price. They include damage to the natural and built environment, such as effects of air pollution on health, buildings, crops, forests and global warming; occupational disease and accidents; and reduced amenity from visual intrusion of plant or emissions of noise. Traditional economic assessment of fuel cycles has tended to ignore these effects."[1] As an example pollution emitted by fossil fuel fired power plants during power generation is known to result in harm to both people and the environment. In addition, upstream and downstream externalities, associated with securing fuel and waste disposal respectively, are generally not included in power or fuel costs. To the extent that the ultimate consumer of these products does not pay these environmental costs, nor compensates people for harm done to them, they do not face the full cost of the services they purchase (*i.e.*, implicitly their energy use is being subsidised) and thus resulting in inefficient allocation of energy resources.

Environmental externalities of energy production and consumption irrespective of production technologies can be broadly summed up in two categories of costs, first, the damage caused to health and the environment by emissions of pollutants having local and/or regional impacts and the second, costs resulting from the impact of climate change attributable to emissions of greenhouse gases. The major source of environment externalities in fossil fuel based power generation is environmental damage from fossil fuels extraction and use. In case of renewable energy sources it tends to be during equipment manufacturing stages and in some of the renewable categories at end use point and fuel use. For example if biomass is being used from forest with out equivalent regeneration, it would lead to deforestation. Similarly the wind energy, if not properly cited, may cause noise, result in bird mortalities, and despoil the aesthetics of landscapes.

5.1 Renewable Energy and External Cost

Renewable energy is usually more environmentally friendly than alternative energy sources and the environmental impacts of renewable energy are site specific. A life cycle analysis (LCA) based upon accounting of all energy and material flows, from "cradle to grave," associated with a system or process is the most appropriate approach in the context of valuation of energy generation technologies as it would analyse the site-specific environmental impact of fuel extraction, transportation and preparation of fuels and other inputs, plant construction, plant operation/fuel combustion, waste disposal, and plant decommissioning. Thus it encompasses all segments including upstream and downstream processes and consequently permits an overall comparison of short—and long-term environmental implications of renewable energy technologies. Further, unlike fossil fuel technologies, the efficiency of renewable technologies is generally very site-specific. Thus, comparisons should be made on the basis of "optimal conditions" costs, rather than the full range that may incorporate old technologies and inappropriate siting decisions. Photovoltaics is generally "delivered" as distributed electricity. Thus its cost should be compared with "delivered" (*i.e.*, inclusive of transmission and distribution costs) electricity from other sources, both renewable and fossil fuel.

There are various estimates about likely life-cycle emissions, from main renewable energy technologies. Table 5.1[2] provides a comparison of life-cycle emissions from renewable energy use with those from fossil fuel plants.

Table 5.1 Life Cycle air Emissions from Electricity Generation (g/kWh)

Energy Source		CO_2	SO_2	NO_x
Fossil Fuel				
Coal	Best Practice	955	11.8	4.3
Oil	Best Practice	818.0	14.2	4.0
Gas	CCGT	430	—	0.5
Diesel	Embedded	772.0	1.6	12.3
Renewable Sources				
Energy crop	Current practice	17-27	0.07-0.16	1.1-2.5
	Future Practice	15-18	0.06-0.08	0.35-0.51
Small hydro		9	0.03	0.07
Solar	PV	98-167	0.20-0.34	0.18-0.30
	Thermal electric	26-38	0.13-0.27	0.06-0.13
Wind		7-9	0.02-0.09	0.02-0.06

Renewable energy technologies generally have relatively high initial capital costs per MW of installed capacity, but low running costs. This many a times result in making renewable technologies financially unattractive compared with traditional fossil fuel derived power using traditional project evaluation techniques based upon the anticipated life of the electricity generating facility. Ideally, for an economic/environmental evaluation, the relevant time frame should be set by the date at which all of the consequences attributable to the project had ceased to exist. However, in the context of CO_2 emissions from fossil fuel power stations this period could exceed 100 years, and in the case of spent-fuel storage for nuclear plants many hundreds of years. Further, it is likely that the value of emission reduction will continue to rise into the future given projected world population growth, economic growth, and the subsequent difficulties in meeting global climate change agreements. In addition, the impacts created on the socio-economic front may also continue for years.

5.2 Extern-E

The European Commission (EC) launched Extern-E (Externalities in Energy) project in collaboration with the US Department of Energy in 1991. The EC and US teams (US later withdrew from the project) jointly developed the conceptual approach and the methodology and shared scientific information for its application to a range of fuel cycles. The main objectives were to apply the methodology to a wide range of different fossil, nuclear and renewable fuel cycles for power generation and energy conservation options. A series of National Implementation Programmes to realize the methodology for reference sites throughout Europe were completed and the methodology was extended to address the evaluation of externalities associated with the use of energy in the transport and domestic sectors. Krewitt has provided a critique of the evolution of the methodologies used in the Extern-E analyses and observed that "It is, however, important to point out that in spite of the existing uncertainties Extern-E results were quite robust with regard to the relative ranking of different electricity generation technologies."[3]

5.2.1 External Costs for Electricity Production in EU

Under Extern-E project, EU has made an estimation of the external cost associated with the then electricity generation technologies. Although the methodology has been applied to a wide range of fuels, different technologies, the results are country specific and can not be extrapolated to derive an average global damage figure. As illustration for attaching a cost on global warming a shadow price for reaching the Kyoto reduction targets has been used which has been estimated in the range of euro 18-46/tonne of CO_2.

Table 5.2 External Costs for Electricity Production in the EU (in euro cent/kWh)

Country	*Coal & Lignite*	*Peat*	*Oil*	*Gas*	*Nuclear*	*Biomass*	*Hydro*	*PV*	*Wind*
Austria	—	—	—	1-3	—	2-3	0.1	—	—
Belgium	4-15	—		1-2	0.5	—	—	—	—
Germany	3-6	—	5-8	1-2	0.2	3	—	0.6	0.05
Denmark	4-7	—	—	2-3	—	1	—	—	0.1
Spain	5-8	—	—	1-2	—	3-5	—	—	0.2
Finland	2-4	2-5	—	—	—	1	—	—	—
France	7-10		8-11	2-4	0.3	1	1	—	—
Greece	5-8		3-5	1	—	0-0.8	1	—	0.25
Ireland	6-8	3-4	—	—	—	—	—	—	—
Italy	—	—	3-6	2-3	—	—	0.3	—	—
Netherlands	3-4	—	—	1-2	0.7	0.5		—	—
Norway	—	—	—	1-2	—	0.2	0.2	—	0-0.25
Portugal	4-7	—	—	1-2	—	1-2	0.03	—	—
Sweden	2-4	—	—	—	—	0.3	0-0.7	—	—
United Kingdom	4-7		3-5	1-2	0.25	1	—	—	0.15
EU range	2-15	2-5	3-11	1-4	0.2-0.7	0-5	0-1	0.6	0-0.25

The costs associated with climate change, such as damage from flooding, changes in agriculture patterns and other effects, need to be taken into account. However, there is a lot of uncertainty about the magnitude of such costs, since the ultimate physical impact of climate change has yet to be determined with precision.[4] The quantifiable externalities included in the Extern-E exercise included public health, occupational health, material damage and also global warming. However, these are dynamic numbers and keep on changing with better scientific evidences about the perceived damages. Table 5.2[5] gives cost ranges (eurocent/kWh) for quantifiable external costs associated with the range of electricity generation technologies for countries within the European Union. The ranges are often relatively large, reflecting variations in generation technology and geographic location.

5.3 Internalising Externalities

Most of the developed countries in their efforts to comply with their obligations under the Kyoto Protocol, have initiated measures that put direct or indirect caps on emissions from energy generations systems. Economic instruments such as taxes, mandatory thresholds on the adoption of low emission technologies or practices by power utilities, renewable obligations etc. are being used to offset externality associated with fossil fuel based power generation systems. Renewables obligations are being increasingly adopted by governments around the world. Such legislation obliges electric utilities to use renewable energy sources to meet a specified target percentage of their supply. The aim is to bring renewable energy online quicker than would otherwise happen by providing incentives for renewables generation. However, estimation of costs of damage associated with externalities of fossil fuel combustion is still considered a subjective analysis that is based on specific case studies and assessments. In view of climate change and energy security issues many of the developed country governments have started levying carbon and/or energy taxes, some of which are illustrated in table 5.3.[6]

Table 5.3 Illustrative Taxes in OECD Member Countries Levied on Electricity Consumption

Country	*Tax*	*Tax rate (in Euro/kWh, except where otherwise indicated)*
Austria	Energy tax	0.015
Belgium	Energy fee (low frequency electricity)	0.0013641
Denmark	Duty on CO_2	0.0134
Denmark	Duty on electricity (heating)	0.0673
Finland	Excise on fuels (manufacturing sector)	0.0042073
Germany	Duty on electricity	0.0128
Italy	Tax on electrical energy, state	0.003
Japan	Promotion of power resource development tax	0.0041
Netherlands	Regulatory energy tax	Varying rates depending upon consumption levels between 0.006—0.02
Norway	Tax on consumption of electricity	0.0128
Spain	Tax on electricity	4.864%
Sweden	Energy tax on electricity (households)	0.0214
United Kingdom	Climate Change Levy (ordinary rate)	0.0069
United States	Delaware: Public utilities tax.	4.25% of gross receipts

In the Indian context there has not been any systematic study that quantifies the externalities associated with the energy generation, transmission and consumption in its entirety. Although, estimation of particulates and gases from such activities have been done at various institutions by different agencies and individuals, an exercise that quantifies them to put a monetary value is yet to be undertaken in a form that has wider acceptability. In the Indian context, even with the present level of developments a number of renewable technologies could become financially competitive with fossil fuel generation if estimates of costs of damage and social costs resulting from combustion of fossil fuels are internalized into the price of the resulting output of electricity. This apart, over the next few decades, the costs of renewable technologies are likely to decline as technical progress and economies of scale combine to reduce unit costs. Incorporating environmental externalities explicitly into the electricity tariff today would serve to hasten the process of transition. Internalizing external costs not only creates incentives for renewable technologies but it also provides incentive for technology improvement in fossil fuel based energy domain.

5.4 Climate Change and Renewable Energy: India Context

The Kyoto Protocol establishes reduction targets for participating industrialized countries (as stated in Chapter1) and provides "flexibility mechanisms" to help countries meet their targets in a globally cost-effective manner. The Clean Development Mechanism (CDM), is the only mechanism through which developing country like India can generate carbon credit and developed countries can purchase these credits towards meeting their own targets. CDM has an elaborate set of rules and regulations and is primarily a project-based mechanisms where concrete projects in respective developing countries and countries with economies in transition can generate carbon credits. The Government of India ratified the Kyoto Protocol on 26 August, 2002 to strengthen the global efforts to meet the challenges adversely affecting the eco-system. The Government of India established National CDM Authority (NCDMA) in December 2003, with its secretariate in the Ministry of Environment and Forests.

Due to its vast market potential for renewable energy projects, and a relatively well-developed industrial, financing and business infrastructure, India is perceived as an excellent country for developing CDM projects. As such India has emerged as one of the most favoured destination for CDM projects globally. Renewable energy projects have major share in the Indian CDM project mix.

5.4.1 GHG Emissions in India

India's first National Communication (2004)[7] is the latest official estimate of GHG emissions from India. In the base year 1994, the aggregate emissions from the anthropogenic activities in India totalled 793.49 million tons (MT) of CO_2, 18.08 MT of CH_4 and 0.18 MT of N_2O. In terms of CO_2 equivalent, these emissions amounted to 1,228.5 MT. On the basis of the global warming potential (GWP) indices, CO_2 emissions contributed 65 per cent of total GHGs, CH_4 contributed 31 per cent and 4 per cent of emissions were contributed by N_2O. The sector wise emissions indicate that 743.8 $MTCO_2eq$ of GHGs were emitted from energy sector (61%); 344.5 $MTCO_2eq$ emissions came from the agriculture sector (28%); 102.7 $MTCO_2eq$ were contributed by industrial processes (8%); 23.2 $MTCO_2eq$ were from waste disposal (2%) activities and 14.3 $MTCO_2eq$ were generated from land use, land-use change and the forestry sector (1%). Most of the GHG emissions in the energy sector are from the coal-fired plants in the electric power industry. The per capita CO_2 emissions were 0.87 tonne CO_2 in 1994, 4 per cent of the US per capita CO_2 emissions in 1994, 8 per cent of Germany, 9 per cent of the United Kingdom, 10 per cent of Japan and 23 per cent of the global average.

Energy-related GHG emissions are mainly a result of combustion of fossil fuels. Among the fossil fuels, coal combustion had a dominant share of emissions, amounting to about 475.53 $MTCO_2eq$ GHGs (*i.e.* about 64 per cent of all energy emissions). The non-CO_2 emissions in this category are from biomass burning and fugitive emissions released from coal mining and handling of oil and natural gas systems. An analysis of the distribution of the total CO_2eq emissions across all the sub-components of all energy activities indicates that the major emitters were the energy and transformation industries (47%), constituting mainly electric power generation, industry (20%) and the transport sector (11%). The all India average emissions from power generation is around 0.84 tCO_2/MWh.[8] However, it may be seen that the build margin which is the average emission of the 20 per cent (by net generation) of most recent capacity addition in the grid is only 0.70 tCO_2/MWh which is significantly less than average emission from the grid. This implies that emission per unit generation are reducing over the years thus reducing external costs.

Table 5.4 Weighted Average Emission Factor of Indian Grids for 2004-05 (in tCO_2/MWh)

	Average	*Simple OM*	*BM*	*CM*
North	0.72	0.98	0.53	0.75
East	1.05	1.18	0.90	1.04
South	0.78	1.00	0.71	0.85
West	0.92	1.01	0.77	0.89
North-East	0.46	0.81	0.10	0.45
India	0.84	1.02	0.70	0.86

Notes: Average is the average emissions of all stations in the grid.

Operating margin (OM) is the average emission from all the stations excluding the low cost/must run sources.

Build margin (BM) is the average emission of the 20 per cent (by net generation) most recent capacity addition in the grid.

Combined margin (CM) is a weighted average of the simple OM and BM.

However, with regard to Indian emission trajectory, the Integrated Energy Policy Report of the Planning Commission has observed that "Since GHG emissions are directly linked to economic activity, India's economic growth will necessarily involve increase in GHG emissions from the current extremely low levels. Any constraints on the emissions of GHG by India, whether direct, by way of emissions targets, or indirect, will reduce growth rates, and impair pollution abatement efforts." It is evident from the likely manifold increase in energy consumption is next 25 years as estimated in the said report.

5.4.2 GHG Mitigation Potential

Given the predominance of the energy sector, the areas have been identified for mitigating GHG emissions include: (a) Electric power generation; (b) Renewable energy; (c) Industry—energy efficiency; (d) Transport; and (e) Municipal waste. The potential reduction in GHG emissions up to the year 2012 (period 2004–2012) has been estimated at 417 $MTCO_2$eq from Power generation (319 $MTCO_2$eq), Transport (57 $MTCO_2$eq) and industries (41MT CO_2eq). GHG mitigation potential of 319 $MTCO_2$eq through 2012 can be realized through fossil fuel utilization improvement (102 $MTCO_2$eq), municipal solid waste (MSW) to energy (63 $MTCO_2$eq) and renewable energy technology (154 $MTCO_2$eq).

5.4.3 CDM Projects in Renewable Energy

CDM has become smaller than the originally envisioned size, primarily because of the rejection of the Kyoto Protocol by the United States, which is the largest emitter of GHGs in the world, accounting for over a third of the CO_2 emitted by Annex 1 countries. The existence of huge amounts of "hot air", mainly from the Russian Federation and the rules concerning the inclusion of sinks in the Marrakesh Accords have further reduced the CDM market. However, the CDM and the emerging carbon market in general have potential to contribute to the financial viability of renewable energy projects, although not necessarily making them fully viable. The World Bank notes that the impact on a project's financial Internal Rate of Return (IRR) with revenue from selling emission reductions will vary depending on the technology and type of GHG being reduced (according to its global warming potential).

Renewable energy sources are that they avoid the air pollution emissions from fossil fuels and reduce GHG emissions from the energy sector, although at the present level on a modest scale. Most long-term energy projections show that renewable energy will play a major role in the global energy supply in the second half of the century, with capacity increasing gradually in the first three decades. The domination of fossil fuels in current energy supplies will not automatically change in the next decades unless renewable energy becomes more cost competitive, or prices for oil and gas remain or surpass current levels.

Renewable energy has also been contributing in mitigating climate change. By January 2008, with 65 per cent of the Indian CDM pipeline renewable energy projects constitutes largest share. At the global level, Indian CDM pipeline constitutes 28.3 per cent of total projects.[9] The details are given in Tables 5.5 and 5.6.

Table 5.5 Indian CDM Project Status

CDM Projects	*Number*	*Annual CER (in 1000)*	*CERs till 2012 (in 1000)*
At Validation stage	460	25646	141936
Requested Registration stage	53	3498	17295
Registered	302	28680	201845
Total	815	57824	361076
% of global CDM project	28.3		14.9

Table 5.6 Indian CDM Project Profile

Category	*Number*
Agriculture	1
Biogas	25
Biomass energy	263
Cement	21
Energy distribution	1
Energy Efficiency in Industry	101
Energy Efficiency own generation	90
Energy Efficiency Service	3
Energy Efficiency Supply Side	15
Fossil Fuel Switch	30
Fugitive	11
HFC	5
Hydro	79
Land fill	8
Reforestation	4
Solar	5
Transport	3
Wind	150
Total	815

5.4.4 Future Prospects for CDM Projects

National renewable energy plans offer ample opportunity for CDM projects with annual capacity addition of: Around 2500 MW renewable power, 1 million Sq m solar collector area,

100,000 biogas plants, 100,000 Solar lights, Wind pumps etc. Technological innovations such as biogas for transport application offer new areas for project development. Further, CDM Executive Board in its 32nd meeting held in June 2007 has approved "Guidance on the registration of project activities under a programme of activities as a single CDM project activity" and the "Procedures for registration of a programme of activities (PoAs) as a single CDM project activity and issuance of certified emission reductions for a programme of activities". In the Indian context, the immediate candidate PoAs could be Programmes for Solar Water Heating, Solar Cooking Family Type Biogas Plants, Cookstove, Biomass Applications in Industry, village Electrification and many more.

5.4.5 Barriers for Large CDM Portfolio

Although there has been significant progress in developing CDM project portfolio in renewable energy in the country, there are some barriers in the operation of the CDM which needs to be looked into if the CDM is to enable significant growth of the renewable energy market in the country. These include the following:

(1) Historically, the CDM under the Kyoto Protocol was intended to be, inter alia, a vehicle for investment in and transfer of renewable energy technologies. It still remains to be realized. Technology plays a central role in addressing climate change. An effective mechanism on technology cooperation and transfer needs to be established to promote R&D, deployment and transfer of renewable energy technologies. It is crucial to remove various obstacles to, and provide necessary incentives for technology cooperation and transfer.

(2) Renewable energy projects are at a comparative disadvantage in the CDM compared to projects which reduce other types of greenhouse gases. The volume of CERs from renewable energy projects is much smaller per unit of output than the volumes created by projects

which abate other greenhouse gases such as nitrous oxide or HFC. As such, the equipment cost of most renewable energy projects is significantly higher per CER than the cost of other types of potential CDM projects.

(3) CER purchasers are generally committing themselves to pay for CERs upon delivery, rather than provide financial support for the project. A smooth implementation of the CDM in India, requires mobilising and establishing a strong capacity initiative in the financial sector. Although some of the major financing institutions have already initiated the process, the institutional set-up in the Indian finance sector require to be further geared up for realizing carbon opportunities in renewable projects. Generally perceived risks include uncertain revenues due to volatile CER markets, uncertain costs for CDM project development, unreliable crediting lifetime for the project, and a variable discount rate. This apart, issues such as perceived regulatory, the higher level of technology risk involved in renewable energy projects have also posed barriers to attracting significant renewable energy project finance.

(4) The transaction costs for developing small-scale renewable energy CDM projects are comparatively very high.

(5) Although now renewable energy projects which are implemented as part of government policies are eligible under the CDM, the additionality issue still remains quite hazy.

(6) There exists an inherent uncertainty about rule and modalities to be followed in the post Kyoto regime. Although the general trends suggest that carbon trading will remain central to any post Kyoto agreement, there is still much speculation in this regard.

Despite the encouraging response of Indian entrepreneurs to the CDM, there remains a large slope across different sectors. Besides there as a need to explore several recent provisions of the CDM such as funding of programmes. The Conference of the Parties (COP) to the VN Framework Convention on Climate Change in its 13th meeting held at Bali, Indonesia in December 2007, requested the CDM Executive Board to approve, at its first meeting in 20C8, the simplified methodologies for "Switch from non-renewable biomass for thermal application by the user" and "Energy Efficiency measures in thermal applications of non-renewable biomass". Approval of there methodologies would have a positive impact on Indian CDM project portfolio and many more projects in biomass would be eligible under CDM.

NOTES

1. http://externe.jrc.es/overview.html as accessed on 12 February, 2007.
2. IEA (1998) Benign Energy? The Environmental Implications of Renewables http://www.iea.org/textbase/nppdf/free/1990/benign1998.pdf as accessed on 12 February 2007 and ETSU Report no R-88 "full Fuel Cycle Atmospheric Emissions and Global warming Impacts from UK Electricity Generation" HMSO, London, as quoted in the first reference.
3. Wolfram Krewitt; External costs of energy do the answers match the questions? Looking back at 10 years of Extern-E; Energy Policy 30 (2002) 839–848.
4. The report from Nicolous Stern (2006) "Economics of Climate Change" has made broad quantification of the cost on global economy due to likely climate change.
5. European Commission (2003): External Costs; Research results on socio-environmental damages due to electricity and transport.
6. ECD (2003): *Environmentally Related Taxes Database.*
7. India's Initial National Communication to the UN Framework Convention on Climate Change (UNFCCC), 2004.
8. Central Electricity Authority, India (2006) CO_2 Baseline Database for the Indian Power Sector.
9. UNEP Risoe CDM/JI Pipeline Analysis and Database http://www.cdmpipeline.org/ as on 8 January, 2008.

Chapter 6

Strategy for Renewable Energy Development in India

The renewable energy can contribute significantly in meeting the Indian energy demand. It can enhance diversity in energy supply markets, contribute to long-term sustainable energy supplies, and reduce local and global atmospheric emissions. Renewable energy can also provide commercially attractive options to meet specific needs for energy services, particularly in remote and rural areas, create new employment opportunities, and offer opportunities to manufacture much of the equipment locally. However, the question is what is time frame within which renewable energy technologies would reach a level when these could be termed as convenient, affordable and financially and economically viable? Obviously at that stage no public interventions may perhaps be required and these technologies would compete with the existing baseline energy technologies. The task in hand is how to reach at that stage in shortest time frame.

In the aforesaid context, any strategy and policy for renewable energy in India should provide a long-term strategic vision in its perspective for role of renewable energy in low carbon growth, where higher resources productivity will contribute to higher long-term economic growth rates, coupled with energy security, lower fuel price, better living standards, new business opportunities within the country and at the global level and creation of new employment avenues. The elements of such a vision would invariably include research and development, policies and programmes for renewable energy application, development of institutes working in various aspects of renewable energy, increased industry participation and availability of finances for renewable energy projects and well directed international cooperation.

6.1 India's R&D: Status and Trends

The national investment on R&D activities was Rs.180.0 billion in 2002-03 which was equivalent to around 0.80 per cent of Gross National Product for the year. [1] In terms of Sector-wise percentage share of national central government's share was 62.6 per cent, followed by state governments 8.5 per cent, higher education 4.1 per cent, public sector industries 4.5 per cent and private sector industries 20.3 per cent. The compound annual rate of growth of R&D expenditure at current prices for the period 1998-99 to 2002-03 works out to be 9.6 per cent and the rate of growth of R&D expenditure at constant prices for the same period was 5.7 per cent. The percentage share of R&D expenditure on energy was 6.0 per cent of total expenditure.

6.2 R&D In Renewables

Central government including public sector has accounted for 67.1 per cent of the total national R&D expenditure during 2002-03. At the national level, major scientific agencies have shared 56.0 per cent of the national R&D expenditure. The MNRE share (central Ministry for Renewables) of R&D expenditure among the major scientific agencies was only 0.1 per cent. In absolute terms it translates into around Rs. 100 million[2] (around 2 per cent of MNRE budget for the year). Year-wise MNRE expenditure on R&D has been stagnating in real terms if inflation is factored in. [3] This expenditure is mainly on specialized institutes such as Solar Energy Centre, Gurgaon; Centre for Wind Energy Technology, Chennai, National Institute for Renewable Energy, Jalandhar and also for funding extramural R&D projects.

Table 6.1 MNRE Research and Development Expenditure

(*in Rs. million*)

Year	*Expenditure*
1994-95	91.9
1995-96	69.7
1996-97	45.7
1997-98	57.6
1998-99	89.5
2002-03	100.0

Limited information is available about development objective wise R&D extramural projects approved for funding by MNRE over a time frame. In SPV and biomass gasification area the sanctioned projects amounted to around Rs. 100 million each respectively. [4]

Table 6.2 MNRE Extramural Projects

Area	*Period onwards-2004*	*No of projects*	*Sanctioned amount (Rs. in million)*
Solar Photovoltaics	1991	25	109.2
Biomass gasifiers	1994	48	94.7
Small SPV/wind energy systems	1991	5	3.6
Biogas	1996	24	33.33

Note: There are extramural projects in other renewable areas, this table has been compiled based on information available on MNRE website, which can not be taken as complete.

Extramural projects in renewable energy are also being sanctioned by departments other than MNRE, such as Department of Secondary and Higher Education, Indian Council for Agricultural Research, Department of Science and Technology, Defence Research and Development Organization and Department of Bio-Technology. In 2004-05, 16 such projects, mainly in SPV and biofuels were approved for an estimated funding of Rs. 31.4 million.[5]

The Technology Development Board (TDB) is a statutory body under DST, providing loan, equity and financial assistance to industry and institution to promote development and commercialization of indigenous technology and adaptation of imported technology for wider application. It has provided a loan of around Rs. 500 million for 5 projects in the area of electric vehicles and power generation from waste. [6]

The Technology Information, Forecasting & Assessment Council (TIFAC)[7] has undertaken technology assessment in the areas of fuel pellets from municipal solid waste, fuel cells, energy tower technology, geothermal energy and bioenegy.

6.2.1 Renewable Research Retrospection

There could be no two opinions that R&D in new and renewable technologies is the key to success in energy transition. After the early 1990s, the renewable energy research has received only marginal attention, apparently due to little directed research focus at the government end and moving towards market based approach seeking industry participation. The industry was not ready to provide funds for R&D in renewables. Even today the industries contribution to R&D in renewables is very low. The renewable research was without explicit consideration as to how a particular project fits into overall strategic goals of the respective technology development. This unfortunately has been the case for R&D projects across the board with most funding institutions. Those funded such research. Much of the research could be characterized as development or technology acquisition and upgrading, *i.e.*, mostly targeted towards incremental product innovation and improved manufacturing or plant operations and performance. There are some exceptions, though—the biomass gasification programme solar PV in 1990s and biogas technology, for example, stand out for its range of indigenous technological development, as these technologies did not find favour in developed countries for the kind of applications suitable in the Indian and developing countries context.

As such, government efforts were restricted to repeating research priority areas with out implementing any strategic plan for renewable technology development in relation to the country's needs and the efforts made were in piecemeal that do not necessarily have synergy between inter and intra technology challenges facing the country. The contribution of academic and other research institutions in renewable research could not increase mainly due to stagnation in research funding from government, which is the primary source of research funding in the country. Further basic research is really limited and most of the research activities are of applied nature.

Though MNRE research policy envisages larger industry involvement, there has been little industry participation. The private sector generally has little incentive to engage in research, or even upgrade its technologies to any reasonable degree. Its research focus was more on solving production problems through incremental efforts rather than new process/product development, and there was little interaction with research community. Further, renewable energy industry, particularly in renewable power area, mostly have one or another kind of collaboration with European and/or US companies, primarily due the technology edge. Such joint ventures have either no or very little research expenditures. Their focus is more on manufacturing and acquiring technology simply from the parent company, which could be perhaps in line with the Government's Technology Policy Statement formulated in 1983 with the basic objective of developing indigenous technology and ensuring efficient absorption and adaptation of imported technology appropriate to national priorities and availability of resources. [8] This was further augmented in the Science and Technology Policy 2003 which has, among others, a stated objective "To encourage research and innovation in areas of relevance for the economy and society, particularly by promoting close and productive interaction between private and public institutions in science and technology. Sectors such as ... energy including renewable energy...would be accorded highest priority."[9]

6.3 Current Governmental R&D Thrust

6.3.1 Integrated Energy Policy Report Recommendations[10]

In its Integrated Energy Policy Report, Planning Commission of India has recommended for setting up of a National Energy Fund (NEF), with an allocation of Rs. 1000 crore for the first year, to finance energy R&D, excluding atomic energy. In new and renewable energy area, the following has been recommended:

- *Bio-energy*: A mission to cover three distinct areas related to bio-energy, viz, (i) Biodiesel from non-edible oils such as Jatropha and Karanj; (ii) Cellulosic ethanol; and (iii) energy plantations should be mounted. In view of the promises offered by production of cellulosic ethanol a separate mission for R&D in this emerging energy source needs to be mounted. Alternate routes to alcohol such as sweet sorghum should also be promoted.
- *Solar*: Given that solar energy is one of our major energy sources and the only renewable energy source with sufficient potential to meet almost all energy needs, high priority to development of solar technology for large-scale deployment should be given. Accordingly a technology mission should be mounted to break barriers to wider use of solar thermal and for bringing down the cost of solar photovoltaic by a factor of five as soon as possible.
- *Hydrogen*: In view of its emerging prospects a mission covering all aspects of hydrogen production, storage, transport, deployment and use, can be mounted.
- *Wind*: Off-shore wind potential needs to be tapped. Wind mill manufacturers to take lead in delivering a time-bound programme and the Government of India may provide fiscal incentives.
- *Community bio-gas plants*: Involvement of individual and community entrepreneurs should be encouraged to experiment with alternate strategies by providing land and finance. Tax rebates to be provided on a graduated scale based on actual outcomes in the field.

6.3.2 Report of the Working Group on R&D for Energy

In the report of the Working Group on R&D for Energy Sector for formulation of XIth Five Year Plan (2007-2012) has observed that "there is no 'silver bullet' and several parallel paths have to be pursued to fulfil the increasing demand for energy for continued economic development. This report has culled out specific R&D areas for various renewable energy technologies. [11] However, there is no mention as to how renewable energy R&D goals are to be achieved and what kind of strategy would be adopted for the same. In addition, there is no mention of international efforts and synergies that are required to remain at the globally competitive R&D levels and to achieve breakthroughs.

6.4 Logical Framework for R&D Strategy in Renewables

R&D framework for renewable can not work in isolation. It will have to complement its efforts, resources (institutional and financial) with the existing R&D framework in the country and also leverage from the international R&D efforts. India has good basic research, an efficient scientific and technical infrastructure and the need is to have innovative approaches to involve larger number of institutions and increase research base in renewables on a sustained basis.

The ultimate R&D objective should be: (a) to make a contribution to the achievement of the India's aims for the expansion of renewable energies, and (b) to further improve the competitiveness of Indian companies and, in this way, to create and secure jobs in India and at the global levels. For achieving these aims the R&D has to be directed towards: (a) to further reduce the costs of renewable energies; (b) to optimise energy systems against the background of renewable energies' growing market share overall; and (c) to ensure the renewable energies sector is expanded in ways that are compatible with the demands of environmental protection and nature conservation.

Ideally, R&D for renewable energy technologies should be viewed as a task for industry that in the present context is implementing most of the renewable energy projects with specific incentives from the government in one or another form. Industry has better incentives for effective R&D as it is the main beneficiary of research findings and better understands the strengths and weaknesses of the technologies. Under competitive circumstances it can act in a prudent manner. In this scenario the role of the government should necessarily be limited to creating conducive atmosphere for innovation and technical progress.

At this stage it would be important to examine why technologies need to the developed within the country when these are available globally or huge funds are being utilized to further develop these technologies in other parts of the world. The explanations could be:

(a) Energy is of strategic importance to any country and government can not shed away with its responsibility to the market forces on the pretext of open economy. Such an approach would neither add value to national competitiveness nor shield the country from changes in international geo-political situation, of which energy is of prime importance.
(b) There is a sense of urgency to achieve energy transition towards cleaner, less and ultimately non-carbon based sources. Here lies an opportunity for the country to remain at the upfront development level and be in position to cater to the global renewable energy technological needs in future, which to an extent is being done by the developed countries, at present;
(c) Development of globally competitive renewable energy industry which can take up business at the global level, adds to country economy in various ways from economic benefit to increased employment opportunities.

In order to put new processes and products in the market the industry has to play its part. However, limitations of present renewable energy products and services can not be addressed through financial instrument only and R&D would need to be supported. This is mainly because most of the Indian industry has not reached a stage where it can comprehend the long time frames for developments in renewable energy technologies from concept invention up to commercial application and these are not in their planning and cost accounting horizons. This apart, the high R&D risks of renewable energy technologies cannot be covered by the market. As such there is a specific need for selective government promotion of R&D of renewable energy technologies, particularly in scenario where industry has not reached to a level where it can absorb the technological risk and factor in the R&D investment over a time-horizon. This sets the guiding principal that government will have to remain at the centre of R&D efforts in renewable energy for achieving national strategic goal of energy security and in longer run energy independence.

Thus at the logical frame the government will have to fund R&D efforts. If so then there must be some definite goals. The R&D funding goals could be articulated as firm contribution to help accelerate the transition to a sustainable energy and increase the proportion of renewable energies in covering primary energy demand.

The first and foremost task is to set specific goals in respect of each of the technologies, examine gaps and/or research needs which should be coupled with focussed, continuous and flexible funding mechanism. Further, support should have a long-term framework, which provides R&D institutions and industry long-term planning perspective for their own research and development work. It is fact that R&D outcome necessarily may not be on the anticipated lines, therefore the R&D support should have necessary flexibility to augment R&D efforts if possibility of break through is apparent during the course of implementation of the project or otherwise absorb failures.

As said earlier R&D can not be attempted in isolation and therefore, coordination among academic institutions, research institutions and industry would be crucial in making a R&D project success and delivering the product in market. Hence, R&D funding support would need to be developed and implemented in close coordination between all the stakeholders as elucidated above.

In Indian context, Government resources are scarce and therefore R&D funding should be planned in such a way that it utilizes minimum resources. To an extent if possible R&D should be organized on competitive basis within the defined boundary conditions through 'request for proposal' approaches. However, while doing so the R&D level should not be compromised and regular evaluation and continuous optimization of structures, regulations and decision-making processes would need to be undertaken.

6.4.1 R&D Framework

R&D needs in renewable energy can be divided in two broad categories. The first are the activities that are aimed at improving products and practical application of the technology. The second are those which foresee long-term technology development needs and create basic framework to achieve such a change. These two distinctive R&D segments would require different instruments for financing. There could be two instruments, viz, institutional funding for basic research and strategic project funding. The institutional funding is not aimed at individual projects, but rather at enhancing the expertise and capability of research establishments for their long-term strategic orientation in the energy research scene. This type of funding tends to be oriented to basic research and typically addresses issues that are best dealt with long term perspective due to their complexity, magnitude and need for specific research equipment. The strategic project funding may be of limited-term closely defined research projects for development of products and systems, which tend to be application—and market-oriented, and whose first practical applications are to be demonstrated. The two instruments complement each other and can be applied together.

Ideally, strategic project funding should address the needs of companies, research institutes and universities in the form of grant for research projects that cannot be realized on the market or for projects that intend to demonstrate the first practical application of new or improved renewable energy technologies. Such funding should be treated as an instrument for supporting projects with a defined topic and time frame involving a high risk which are of national interest. It should only be applied in cases where the market itself will not be able to provide the new technological developments within the foreseeable future.

Strategic project funding could be in the form of collaborative research, in which the universities and research institutes cooperate with commercial companies at the pre-competitive stage in order to improve existing energy technologies or develop new energy technologies by an inter-disciplinary division of labour in dealing with problems of a complex nature that can only be solved in the long term.

Long-term technology development needs which includes basic research of the concepts that shows some initial promise but far away from the product development stage could better be dealt by institutional funding. This require continuous support to institutions for creating groups that keep on working in identified R&D areas on long-term basis in a mission mode and act as centre of excellence for the respective technology areas. However, centre of excellence should not mean that specific institutions be set up by government for the purpose, it could better be within an existing institution, which provides a natural benefit of existing core group and availability of academics and experts from other stream. Continued support to IACS's solar thin film and photovoltaics facilities and funding to IISC's biomass research are the examples of such funding.

6.4.2 R&D Project Appraisal Mechanism

A project management mechanism that examine every project submitted with respect to its innovation content as well as the expertise and the creditworthiness of the applicant, needs to be created. This mechanism must result in assessing the possible contribution that the project may make to the national renewable energy R&D programme. The level of funding should be determined by the scientific and technological risk and the national interest attached to the project. In principle the level of funding should be determined by the principle of the economic and thrifty handling of public funds.

The institutions identified for working in the basic research areas through institutional funding should be involved in project funding exercise. One of the major responsibilities of the identified institution should be to develop linkages with other institutions and industry for developing project proposals in the identified areas. It should be involved in the projects from the time the applicant first makes contact through the application procedure, evaluating the application, decision on funding, disbursement of funds, efficiency control and final accounting up to exploitation of the results. Speed and transparency must be the guiding principle for such a mechanism.

6.4.3 Specific Issues for Future R&D

Identification of R&D areas alongwith organizations that could take up the activity is key to a successful R&D. In doing so taking lessons from the successful models as adopted by EU, Germany, Japan and USA would be very useful. Here-in-under are some of the major technology specific research areas.

6.4.3.1 Wind Energy

6.4.3.1.1 Near-Term

The overarching focus of future research and development work on wind technology should be directed towards cost reduction and increasing the output of wind power systems, as well as greater automation in the production and construction of wind power systems. Technological development and cost reduction of wind power are directly related to turbine size and with increase in average turbine size the investment cost per MW of turbines decreases and accordingly per unit cost of energy also decreases. This apart, other research priorities could be use of improved materials with higher strength to mass ratios and reduction in number of components.

Handling of wind intermittency is another issue. With improved performance prediction through anemometry, terrain calibration and refined power-wind speed curves, better load management could be attained by the utilities.

Standardization of sub-systems and components would increase modularity and would contribute to reduction of cost of system design, engineering and production.

Further development of small wind aero-generators and wind/hybrid systems with very low maintenance requirements and subsequently their standardization would help in meeting off-grid electrification needs for remote locations. In addition, the potential of wind pumping for meeting irrigation needs to be assessed and research efforts for developing designs suiting to field conditions would require to be undertaken. This area needs specific focus.

6.4.3.1.2 Medium-Term

Although the preliminary estimates are not very encouraging, off-shore wind power needs attention for long term wind development goals. It would inter alia require a coordinated approach first for potential estimation, design development through appropriate international collaborations and thereafter setting up some demonstration projects to prove the viability so that private sector could take a lead thereafter.

6.4.3.2 *Small Hydro Power*

6.4.3.2.1 Near-Term

The engineering of hydroelectric plants is largely mature. There have been a number of improvisations for reducing cost of civil works and hydro turbine. As such R&D priorities are of operational nature with the focus on development of best practice manual for setting up of a KW size small hydro power system and standardized operation and management schedule. Another important area could be grid interfacing issues of small hydro power systems.

Creation of small scale decentralized manufacturing, repair and maintenance base would help in dissemination of small hydro systems in a large numbers. It will also result in better maintenance of existing systems in the respective areas and also accelerated installation of systems. Techno-economic studies for setting up of such bases would need to be taken up.

Direct mechanical application of hydro-power has remained a comparatively neglected R&D area. Best practices to set up such systems needs to be developed, along with techno-economic analysis for setting up such systems. This would enable large scale dissemination of these systems.

6.4.3.3 Bio Energy

6.4.3.3.1 Near-Term

- *Biomass Gasification*: Further investigation of the special combustion and gasification properties of biomass is an important area for further R&D. This could facilitate development of optimal technologies for respective biomass utilization. Areas that needs to be addressed, include system design, gas cleaning and efficiency of generating electricity from biomass via the technology of gas engines or gas turbines to that of traditional methods.
- *Biodiesel*: Development of processes, fuel standards and quality management.
- *Biogas*: R&D in biogas generation has two distinct aims, the first, to develop more cost effective, easy to operate, maintain and reliable systems for use of common masses for cooking, lighting and motive power purposes; and the second for commercial applications through improving conversion efficiencies under controlled condition and developing packages for alternate biogas uses for transport and other purposes.

6.4.3.3.2 Medium Term

- *Bio-hydrogen*: Extraction of hydrogen from biomass economically is another important R&D area. At present there is no standardized method for extraction of hydrogen from the biomass. In view of the work going in hydrogen as a long-term option for emission-free energy utilization, development of biomass based hydrogen generation systems is an important area for further investigation.

- *Liquid fuels from Biomass*: Generation of liquid fuels from biomass is an important R&D area. Production of synthetic vehicle fuels by the gasification of biomass and production of Fischer-Tropsch fuels is being investigated. However, the procedure has yet not achieved the maturity necessary for practical application. As there is even not a single demonstration project in the country based on this technology. There is an urgent need for development of this area through coordinated approach.

Biomass-based ethanol technologies are still evolving and the commercialization of this technology has to overcome various bottlenecks. Keeping this perspective in view, bioethanol technologies needs to be analyzed in terms of feedstock availability, pretreatment strategies, efficient hydrolytic agents, availability of recombinant ethanologens and process economics with a focus on Indian research efforts.

Extensive research in use of lignocellulosic biomass for ethanol production is being carried out globally. Some groups in India have also initiated work in this direction. The main R&D need in this area is optimization of fermentation parameters for high conversion rate of lignocellulose into ethanol.

6.4.3.4 Solar Photovoltaics

6.4.3.4.1 Near-Term

A mission approach would be required to address the multifarious dimensions of solar PV programme which encompasses a wide spectrum of issues, responsibilities and activities such as: (a) fundamental research; (b) advanced materials and devices; (c) systems technology development; (d) performance and reliability testing; and (e) manufacturing and deployment. The mission objective could be to develop Indian photovoltaic research, technology, manufacturing, applications, market, and policy aimed at solar PV cost reduction, creating indigenous manufacturing and deployment for niche applications till such time solar PV becomes cost-competitive with mainstream energy options with the ultimate objective of contributing to national energy security and economic development.

6.4.3.4.2 Medium-Term

Launching of a National Coordinated Research on Solar PV (NCRPV) on different aspect of solar PV ranging from basic research, material development to efficiency improvement. The NCRPV tasks could include basic research on electronic materials and devices, processing science, performance advanced research in tandem or multi-junction solar cells, measurement and characterization; operational research on crystalline silicon, thin films, advanced PV concepts relating to organic, dye-sensitised, quantum/third-generation cells, building and other integration concepts; systems technology development through systems modelling and analysis, benchmarking and validation, system performance and standards; and development of procedures for type testing and development of standards.

6.4.3.5 Solar Thermal Power

6.4.3.5.1 Medium-Term

In view of future potential and prospects there is a need to focus on solar thermal power issues and develop a road map for developing technology. Demonstrating the solar thermal power project would help establishing operational viability under Indian conditions and would result in follow-up investments by the private sector both in the manufacture of the solar field components and in larger solar stations within India. This ultimately will determine if this technology is replicable in other parts of India. As such a coordinated project for research and development of concentrating systems covering various solar thermal power technologies and also development of the engine would need to be launched.

Specific R&D areas that needs to be looked into would include development of heat transfer medium, receiver tubes, parabolic reflectors, tracking system, large area solar dish, stirling engine for small power applications, etc.

6.4.3.6 Low and Medium Temperature Solar Systems

6.4.3.6.1 Near-Term

On technology development front specific projects for development of advanced solar collectors with higher optical efficiency for flat plate collectors and more efficient and cost effective designs of solar cookers to suit different cooking habits of the people and studies related to storage of heat for use during off-sunshine hours would be desirable. In addition, standardized and cost effective designs of solar air heating and drying systems for different agricultural products is required.

Another important area for research is the development of cooling of systems for space cooling including cold storages.

6.4.3.7 Hydrogen

6.4.3.7.1 Medium-Term

R&D in Hydrogen is at a nascent stage in the country. In view of the technologies that have already been developed in other parts of the world, the efforts made so far needs consolidation and new approaches would need to be devised. Future challenges include lowering cost of hydrogen substantially and improve production rates from different methods, development of compact and inexpensive storage capacity, establishment of hydrogen network, development of hydrogen fuelled IC engine with costs comparable to existing petroleum based IC engines and cost reduction and efficiency improvement of different types of fuel cell systems. This would ultimately require launching a mission for the purpose.

Specific R&D activities could be standardization and cost reduction of production of hydrogen from natural gas and biomass-derived renewable liquids, advanced renewable photo-electrochemical and biological hydrogen generation technologies, high-temperature thermo-chemical cycles driven by concentrated solar power processes to produce hydrogen, on-board hydrogen storage systems, direct hydrogen fuel cell power system for transportation, distributed generation polymer electrolyte membrane (PEM) fuel cell system operating on natural gas or liquid petroleum gas, electrolyzer powered by a wind turbine, safety procedures etc.

R&D in developing hydrogen-fuelled IC engines for stationary and transport applications would support the early deployment of a hydrogen infrastructure and therefore must be taken up on priority basis. In the long run fuel-cell technologies are expected to take over and thus will have to be looked into from the initial stage from now.

6.4.4 Overarching Technology Development Issues

Pace of technological development is key for larger application of renewables in the country. The spillovers of technological growth could be witnessed in the form of increased business and economic activities by businesses and industry in other countries. Therefore, it is utmost important that industry is encouraged to make investments in technologies aiming at cost reduction and improving performance. This needs to be facilitated through creating conditions, including larger availability of venture capital fund, for increasing manufacturing capabilities within the country.

In order to have better products and reliability, it is important that codes are enacted and standards are set for a range of renewable energy systems and products available in market. In addition, putting in place an appropriate certification mechanism will ensure quality and also ensure reliability.

There is an urgent need to carry out detailed assessments of wind, biomass, geothermal and solar energy resources and fix the total resources volume, technically exploitable volume, and economically dynamic exploitable volume. This will provide a reliable basis for policy and project planning and design.

Development of hybrid renewable systems and products with appropriate storage systems that provides optimum system efficiencies and can absorb intermittency is ideal for off-grid decentralized applications. This area has yet not received enough attention, although it has large application potential.

6.4.5 International R&D Cooperation

The present investment policy regime and transport facilities have made it possible to set up renewable energy R&D facilities in such a location at the global level where they achieve best financial prudence. R&D activities of an industry that operates at global level and movement of technical know-how provides best conditions. The surge for development can be seen within industry. Acquisition of wind R&D facilities in Germany by an Indian wind energy company is a typical example. Otherwise also for achieving excellence in R&D and maintaining competitiveness exchanges between and/or with the scientific and technological professional in an international framework is must. Further. international R&D cooperation avoid any duplication of efforts and make the most efficient possible use of scarce resources. This could be achieved through well defined R&D projects with proper division of labour and responsibilities for specific tasks with equitable financial burden and credit sharing arrangements. Bilateral and to an extent multilateral scientific and technological cooperation agreements could provide framework for such R&D activities, including for an exchange of personnel and setting up demonstration projects. Indian association with IEA's R&D implementing agreements and EU R&D framework could be useful in terms of understanding of the frontier areas of research and getting best out of participation. As a country, India has an unique position as it has the perfect balance in terms of give and take having highly skilled manpower, readiness for financially shared R&D, better understanding of developing country needs and markets, and vision to achieve global competitiveness.

NOTES

1. Department of Science & Technology, 2007.
2. Department of science and Technology: Research and Development Statistics 2004-05 http://www.nstmis-dst.org/ContentsPage.htm as accessed on 15 February, 2007.
3. DST: http://www.nstmis-dst.org/rd_stat/updated/rdstat/table4.htm as accessed on 15 February, 2007.
4. MNRE website http://www.mnes.nic.in/frame.htm?major-prog.htm as accessed on 15 February, 2007.
5. DST (2006): Directory of extramural research & development projects approved for funding by selected government agencies/ Departments during 2004-05. It has also observed that "despite repeated reminders and visits, data was not provided by the Ministry of Non-conventional Energy Sources".
6. Technology Development Board (2007) http://www.tdbindia.org/ben.htm as accessed on 15 February, 2007.
7. TIFAC is a registered society under DST with the main objective of generation of Technology Forecasting/Technology Assessment/ Techno Market Survey documents and enabling technology transfer as well as commercialization. http://www.tifac.org.in as accessed on 15 February, 2007.
8. Department of Science and Technology http://www.nstmis-dst.org/Stp3.htm.
9. Science and Technology Policy 2003, http://www.dst.gov.in/stsysindia/stp2003.htm#c13 as accessed on 27 April, 2007,
10. Planning Commission (2006) http://planningcommission.nic.in/reports/genrep/rep_intengy.pdf) as accessed on 19 February, 2007.
11. Principal Scientific Adviser to the Prime Minister's Office (2006) http://psa.gov.in/energy.pdf as accessed on 19 February, 2007.

Chapter 7

Renewable Energy Policy: Directions for India

The existing policy provisions for renewable energy have already been described in Chapter 4. However, with rapid developments that are taking place across the world the policy framework has to be dynamic so that it captures the changing requirements. The ultimate objective of the renewable energy policy framework is to significantly increase the share of renewable energy sources in India's energy mix. Such a framework should be ideally stable that has long-term mandatory targets with flexibility in target setting across sectors. It should provide for continued efforts to remove barriers to renewable energy application and take into consideration environmental and social aspects. The framework should result in cost-effectiveness of policies, and be compatible with the overall internal energy market.

Renewable energy technologies vary widely in their technological maturity and commercial status. Even with the present level of renewable energy technologies, when some of renewables can compete with conventional fuels even on financial terms there exists a number of tariff and non-tariff barriers that hinder the increase of renewable energy technologies market penetration. As such the Indian energy market has never been competitive and it is unlikely to be fully competitive in the near future. Investments in infrastructure and long-term subsidies for conventional energy along with the failure to internalize external costs and benefits, all act as obstacles to the advancement of renewable energy. Renewable energy is at the take-off stage and businesses, industry, government and customers have a number of issues to address before these technologies could make a real penetration. There is a need to have a policy framework that addresses various barriers.

Government of India is yet to announce a long term policy that takes into account development and dissemination of entire gamut of renewables. This apart, policy should be able to move towards a more level playing field for renewables and an assurance of a stable market. In view of the diversified nature of renewable energy technologies, it would be more appropriate if a combination of policies through correctly defined policy instruments are applied. However, the question remains of how to achieve these goals and what policy mechanisms need to be put in place to move toward the desired outcome. The policy suggestions here-in-under does not purport to offer straight jacket mechanisms, it at best proposes some of the policy measures that are relevant and lead to better harnessing of renewables.

7.1 Incentive Policies

Application-wise renewable energy can broadly be categorized in five distinct groups, viz, Grid-interactive renewable power, Decentralized & stand alone renewable electricity systems, Heat energy for cooking purposes, Process heat for domestic, industrial and commercial purposes, and Liquid biofuels for transport applications. Except renewable power which is governed under the provisions of Electricity Act, 2003 and related Electricity Policy and Electricity Tariff Policy, and have assured market in terms of existing and impending RPS by States, renewable systems for other applications are mainly linked either to annual targets of MNRE and availability of financial incentives. Technology can develop only when markets are projected. Longevity and predictability of policy support are important for overall market success.

A general principle of any sustainable renewable energy policy is that the project developer should be able to recover his project cost within a defined time-frame. Government could help by making clear to whom these costs will be disbursed or in other words the additional cost burden should be spread across all the citizens who benefit from the project, both from the energy as well as the environmental benefits. All types of economic incentives should be time limited through sunset clauses.

In the present policy framework incentives for renewable power are in various forms of tax benefits and capital subsidies. In line with the recommendations of the Energy Policy Committee (2006), phasing out the capital subsidies would help to move renewable power without annual budgetary support. An alternative mechanism for linking incentives to energy generated would spur the development process.

7.2 Tradable Renewable Energy Certificates

Globally, two major and different compensation systems are used in order to promote the generation of electricity from renewable energy sources: (a) Feed-in Tariffs for priority access of renewable energy sources to the grid at guaranteed tariff for a specific and defined time period; (b) Quota obligation and/or green certificates in which government defines quota target and renewable certificates are traded at market price. Additional income is generated from certificate trading and prices for certificates are based on quota target and depend on market variation. Electricity Act, 2003 provides for feed-in-tariffs and Renewable Portfolio Standards and many of the States have already announced feed-in tariff and quota for renewables. However, in absence of a renewable energy trading system effectivity of quota system is uncertain.

In the Indian context, the Electricity Act, 2003 does not give any flexibility to create Renewable Portfolio Standards (RPS) or Quota for renewables. Introduction of tradable renewable energy certificates (REC) could overcome the existing gap that is hindering the application of quota for renewables and thereby create a vibrant market. REC is being applied in many countries and its use is growing the world over. It provides a flexibility to the electricity providers in meeting their RPS requirement. They can generate the necessary amount of renewable electricity themselves; they can purchase it from someone else; buy credits from other providers who have exceeded the standard or from the spot market (as is the case in developed country). They choose the option that is cheapest for them. The REC system if introduced at the national level would allow trading of green attributes seperately from physical electricity and therefore would remove geographical or physical limitations, such as resource availability and/or time scale such as seasonal availbility or financial limitation associated with supply of renewable power. As such operationlizing REC at the national level would be desireable for effectively meeting RPS commitments as enabled under the Electricity Act, 2003. However, while doing so aspects relating to compatability with the existing laws and development of a robust system for validating, registry, verification, transfer and tracking of RECs would need to be looked into. Eventually, REC market would be akin to stock markets, as is the case with carbon credits.

7.3 Internalization of Environmental Externalities

The principle of internalising the environmental externalities of CO_2 emissions and other pollutants resulting from fossil fuel combustion is now globally accepted and many countries have already started internalizing these costs in various ways. This subject has been dealt with in detail in chapter 5, Renewable Energy and Environment. Such an action will result in rise of the cost of power generation based upon fossil fuel combustion and a relative improvement in the competitive position of an increasing range of renewable energy technologies. In other words, the removal of both direct and indirect subsidies to power generation technologies and the appropriate pricing of fossil fuels, to reflect the environmental damage created by their combustion, are essential policy strategies for stimulating the development of renewable energy technologies.

7.4 Manufacturing Volumes and Market Creation

In addition, focused and directed technology development, reduction in the cost of renewable energy facilities and products depends on development of market and creation of volumes. Any government policy should aim at creation of markets that best address the needs at the least economic cost. An increase in manufacturing volume for renewable technologies will result in substantial cost reduction. The level of the cost reduction varies with the type of technology and increased volume of equipment production results from increased market demand. The most effective policies that increase market demand are market-mandated policies and if properly designed and enforced these policies guarantee the demand for technologies that will generate the economies-of-scale in manufacturing needed to bring down costs.

Domestic manufacturing and reducing import dependence is important for a country like India, where market volumes exist and accordingly industry has the appetite and motivation for the same. However, it requires careful policy interventions and planning that allows for the steady increase of component manufacturing. Renewables are increasingly getting indigenous in India, and many a time import is an economic decision. Reduction in import depends on development of indigenous technology or technology transfer. Technology transfer generally include a combination of gradual shifts in import tariffs, incentives for joint ventures, development of equipment standards and testing programme, and standardized criteria for competitive solicitation.

7.5 Institutional Finance

The structure of financial requirements and scale pose a challenge to investment in renewable energy. Renewable energy projects usually carry higher up-front capital costs and lower operational costs than their conventional counterparts. The external financing requirement is therefore high and requires amortization over the entire project lifecycle. This apart, most renewable energy projects are small and therefore have comparatively higher transaction costs in terms of feasibility analysis, legal and engineering fees, consultants, etc.

The annual investment of the renewable energy industry in the country has reached a level of over us 5 billion. Over the coming years it is set to increase with the increased level of installations. To obtain the needed investment, renewable energy projects will have to compete with other conventional segments of the energy industry. However, renewable energy is becoming more competitive and investment is growing. Investment in renewable power has already registered a high growth. The renewable energy share of the venture capital market for new technology development is also growing, although more slowly than other environmental technology areas. However, traditional financiers and investors still view renewable energy as high risk and low return option.

An integral element of a project financial structuring is risk management. Linking renewable energy projects with carbon opportunities in the form of CDM projects is one of the new ways of improving project viability through carbon money and also minimizing risks. With impending carbon trading opportunities more and more banks have now started looking at renewable energy projects as a viable financial proportion.

Investment in renewables can be put in three categories. The first is where investments are financially attractive under sound, commercially viable policies. The second address promising new technologies that are not currently financially viable, but investment becomes viable with subsidies. The third category mainly relates to the technologies at advanced research and development stage that could yield commercially viable results within medium term and in the short-term requires venture capital support.

In the Indian context, renewable energy programme is primarily private sector driven. The investment climate for renewables has now changed significantly from the initial years of the programme around two decades ago, when Indian Renewable Energy Development Agency (IREDA), a non-banking financial institution under the Ministry of New and Renewable Energy was meeting major portion of institutional finance needs.

Now mainstream banks have started providing finances for renewable loans.

A major area that needs attention is micro-credit for off-grid renewable energy applications where getting the finance is the real issue. Although a number of initiatives that cater to micro-finance needs in a specific area/region are already underway, there is an immediate need to develop a national framework which ensures refinance to the state and regional banks against the loans given for renewable energy systems. There exists a precedence of NABARD refinancing banks for the loans given by them for biogas plants. In nutshell innovative financing strategies would need to be developed/augmented for: (a) micro credit or micro-lending; (b) financing leaning programmes; (c) vendor credits; (d) targeted project credits; and (e) equity financing/debt financing. Ideally, IREDA should take a lead and develop a diversified institutional finance management approach that takes into consideration future requirements and also involves mainstream financial institutions at all the levels.

7.6 Technology Specific Future Directions

7.6.1 Wind Energy

In recent years, the generation of electricity by wind power has experienced considerable growth in India. The major factors that are responsible for the growth of wind power installations in the country include fiscal incentives in terms of accelerated depreciation and implementation of the provisions of Electricity Act, 2003. At present the market size, through a defined quota, as well as returns through set tariffs, are assured. In the long term, as per the National Tariff Policy there would be competition within same renewable energy technology and subsequently with all renewable energy technologies. With improvement in technology for attaining higher conversion efficiencies for harnessing low wind regimes, identification of new wind sites including off-shore, continuous expansion of grid power electric installations and also increasingly economic viable potential it could be safely assumed that wind power systems will continue to be set up initially in 'greenfield areas' and subsequently in old sites with new and higher efficiency systems for next two decades or so till new renewables offers more competitive solutions.

The players in the Indian wind market have been established over the last decade and consolidated over the last five years. Entry of new players in manufacturing/project development in this market will have to be with new/competitive technology or service. One of the limiting factors in the Indian wind market is captivity syndrome where manufacturers are having dual role of project development. As such a market segment of project developers is yet to be identified. An effective decoupling will have a positive impact in terms of project cost reduction, competition and better transparency and accountability.

7.6.2 Small Hydro Power

Small hydro power plays a very important role in developing local economy and fufilling energy needs of remote areas. Although there has been significant interest of the private sector for developing small hydro projects, this sector is not growing at the desired pace. It is well recognized that SHP systems not only help meet the ever increasing electricity needs, particularly in remote regions, but add significantly in overall socio-economic development of the region. The trend indicates that there has been no acceleration in SHP installations in the country and with the present pace it would take more than 100 years to exploit even the identified site specific SHP potential. This is unacceptable to every one.

In view of terrain inaccessibility and lack of appropriate linkages the investment costs of SHP projects are comparatively high. Further, there exist problems relating to load development as remote hilly areas having high SHP potential usually have low demand for power, thereby limiting its viability as a decentralised power generation option. High investment is required for construction of transmission networks in difficult terrain for transmission of power to areas with higher demand. This apart, lack of long term discharge data of potential sites, lack of coordination among planning and implementing agencies, delays in the clearances and allotment of private sector projects are other common reasons for very slow growth of SHP in the country. In the case of stand-alone systems which involve local participation, success of small hydro development depends to a large extent on local capacity building programmes, setting up of appropriate institutional arrangements for demonstration, training and awareness programmes that help in technology adaptation and maintenance.

The logical policy aim should be to harness the existing SHP potential in shortest possible time frame. However, the present policies for SHP development are more regulatory in nature than being developmental. While on the one hand it focuses on SHP projects conforming to the international standards, on the other hand it does not create an atmosphere for involvement of civil society in a larger way. This process has alienated SHP technology from the common man and has remained in the hands of a few with regime of inspection and supervision. As such a new vision and approach that makes SHP a mass movement with sustainable economic returns is desirable.

Let us see it in another perspective. Small hydro has been common man's technology for centuries and throughout Himalayan region hydropower is used for grain grinding. There exists a cadre of traditional people who understand the basics. An approach that provides incentives for creation of SHP manufacturing and generating base at the small scale where manufacturer could double as developer would have far better results not only in terms of added capacity but also better functionality, utility and decentralized applications. Local capacity building and micro citing of systems which not only focuses on power generation capacity but also on mechanical application would be required to achieve the desired results. A step further would be to develop a mechanism that provides for the integration of all SHP systems to the national grid. This would result in better load management and generation of additional revenue stream to the local communities who own such power plants.

7.6.3 Solar Photovoltaics

Indian solar PV technology development efforts has so far focused on crystalline silicon solar technologies. Efforts to develop alternate solar technologies were half hearted and did not produce the desired results. SPV module and system manufacturers are highly dependent on imported silicon wafers. The government has not developed any coordinated programme for development of new solar PV technologies, as is evident from the fact that in 2006 there were only 2 MNRE supported research projects under implementation. It may be recalled that India was one of the first countries to develop solar PV through technological development and also large scale deployment for meeting subsistence energy needs to the rural masses. The National Project for Solar Photovoltaic Demonstration was initiated in 1975 when the technology was too far from becoming competitive.

Solar energy for supplementing grid has recently found support of the Govenment. In December 2007, the Ministry of New and Renewable Energy has announced a scheme to provide financial assistance upto Rs. 12/kwh in case of SPV and Rs. 10/ kwh in case of solar thermal power fed to the electricity grid. The scheme provides for setting up of a maximum capacity of 10 MW in each of States and a maximum of 5 MW per developer. In the initial phase incentives will be provided for 50 MW capacity. In view of rapid developments and future possibilities there is a need to put together experience in this segment and a policy instrument that provides for controlled dissemination as a right policy decision. This would help in generating experience in solar power technologies development thereby reducing costs by creating markets. Although roof top solar systems and building integrated PV systems have been launched successfully in western countries, but this involves a large number of generators to deal with and would call for interaction with numerous distribution companies, which may not be practical option in the Indian context to start with. Therefore, the present scheme us in right direction, however, in the longer run, roof top SPV systems would need to be covered.

There exists a significant opportunity for stand-alone solar PV application which offers a viable option for household lighting in rural areas. As per the Census of India (2001), 83.12 million rural households in the country use kerosene as a primary source of lighting, despite having grid in most of these areas. Various studies indicate that on an average 2.0 litres kerosene oil is consumed per month per rural household, thus around 2000 million litre (in 2003-04 there was total consumption of around 10 million tonne) of kerosene oil is being consumed for lighting purposes in the country. For 2004-05 average subsidy on kerosene was Rs. 8.78/litre. Thus total subsidy for kerosene based lighting for 2004-05 would be Rs. 18,000 million (overall subsidy on kerosene in 2004-05 was over Rs. 100,000 million). This subsidy is sufficient to cover around 45 million households with solar PV based home lighting systems by providing an one time subsidy of Rs. 4000/system. Within two years entire rural population of the country using kerosene lighting systems could be covered instead of regular kerosene subsidy. This policy initiative on the other hand will create significant demand for SPV systems within the country resulting in additional industrial activities thus additional employment and economic activity.

7.6.4 Solar Thermal Power

Solar thermal power technologies offer interesting opportunities for India. These technologies can easily be adapted to Indian conditions and most components of solar thermal power plants could be produced indigenously. Success of efforts will depend on to what extent the impact of the initial investment cost is mitigated, operating costs fully recovered, professional management introduced, and infrastructure and equipment support for operation and maintenance made accessible. In this process initial government support will be crucial.

7.6.5 Solar Heat

Solar thermal technologies, particularly solar water heating systems, solar cookers and solar steam generation systems are the most commercialized technologies among renewable energy technologies in India. Large potential exists but high capital cost is a major barrier in their large-scale utilization. These systems do not produce but save power and are still not considered as important as power producing technologies such as wind, solar PV or biomass by the investors. The investment in the sector is minuscule as compared to that in other renewable energy sectors. In this sector large scale capital investment and also introduction of advanced technologies are needed in the country. Few State governments such as Karnataka have given boost to solar energy utilisation in domestic sector by providing a rebate of 0.15 Rs./kWh of electricity consumed to consumers using renewable energy systems to save electricity. This has given boost to domestic solar water heating systems in these states. Such policies will provide further impetus to dissemination of these technologies.

7.6.6 Biomass-based Power

Power projects based on biomass plantations have the potential to offer new avenues of employment through collection, storage, handling and utilization of biomass materials especially in rural areas, promote rural industries and generate rural employment. Besides, energy plantations can become the means to restore deforested and degraded lands in tropical and subtropical regions. By a broad estimate, 20 m ha of wastelands yielding 10 MT/ha/annum of woody biomass giving 4000 k cal/kg with system efficiency of 30 per cent and 75 per cent PLF would be enough to set up 45000 MW biomass based power generation capacity.

Whereas initial government focus was on awareness building programmes through launching demonstration programmes in cooperative/private sugar industries, the focus now is to encourage private participation for setting up power projects. However, various issues need to be addressed to exploit the potential of cogeneration based power generation. Large incremental investment requirements for setting up cogeneration units are a constraint in setting up cogeneration facilities in some of the existing industries. In addition, several sugar mills in the cooperative sector continue to be sick or closed and require special financial support to revive/modernize and adopt cogeneration technologies. Some of the technical barriers faced by the industries are upgradation of existing mills, installation of power generation systems, synchronisation and feeding of electricity to the grid. Sometimes the sugarcane bagasse may be channelled for alternative uses, *e.g.* for paper production. The factors that are likely to shape cogeneration based power are future energy prices, both from cogeneration as well as from competing systems; presence and expansion of those industries from which a high potential for power generation using cogeneration techniques exist; technological developments, etc.

For setting up of large-scale biomass based electricity generation technologies, the most important issue is ensuring a continued and reliable supply of biomass, especially woodfuels. Sustained supply of biomass will require enhanced production of energy crops where critical issues related to land availability, enhancing productivity through technological interventions and other economic operations related to biomass supply will come to the fore. An important issue in this context would be development of waste lands for ensuring reliable supply of biomass so that it does not compete with foodgrain production on main agricultural lands. The delivered cost of fuel, including the cost of biomass and the transportation cost, is an important factor that determines the competitiveness of this form of generation with that from other sources.

7.6.7 Biogas

India's Biogas Programme is going through a crisis phase. In order to revitalize this programme so that its full potential is exploited a complete overhauling of the programme would be desirable. The changed strategic efforts may include adopting cluster approach in selected districts/blocks, based on biogas plant performance over the last two decades; addressing operational and technical glitches by reorienting the research machinery and by encouraging private sector investment in biogas research;

investing in design research to make biogas suitable to different climatic conditions; develop models that take into account local problems like low temperature, less availability of water, alternate feeding material and proper feedstock mix; proper incentive scheme for batch-fed biogas systems, kitchen waste biogas plants and human excreta based biogas systems. NPBD needs reorientation to accommodate developments in the sector and introduce new designs for speedy implementation. Re-assessment of biogas potential at district/block level for all possible biomass resources would help in increasing the coverage and outreach. Finally, the present approach of allocating token targets of around 1 plant in every 6 villages does not create an environment for proper institutional framework nor help in exploiting the potential in the foreseeable time frame. For the dissemination of this technology, volumes are of utmost importance as they are required for creation of effective operation, repair and maintenance services.

Another area that needs attention is participation of private entrepreneurs in setting up of biogas plants. This action needs to be developed that takes into consideration local and regional variations in terms of the biogas market. Parallelly it will also be required to develop a marketing strategy.

In addition to its cooking application, there is a potential of biogas generation in India to make it an alternate fuel for vehicular transport. In order to have more calorific value and better fuel quality, biogas requires enrichment before bottling for storage and mobile purpose, as enriched biogas has. IIT, Delhi has developed a method of biogas enrichment through water scrubbing which is simple, low-cost and suitable technology for enrichment of biogas in rural areas. The designed and fabricated biogas scrubber is able to remove 99 per cent v/v of carbon dioxide present in raw biogas. It has been established that a biogas enrichment and compression system is a profitable venture for rural areas due to availability of large quantity of cattle dung. The system is recommended to establish rural entrepreneurship for the effective utilization of local biomass resources for production of biogas energy in a decentralized manner and sustainable rural development.

7.6.8 Bio-Fuels

In India, the large-scale development of bio-fuels, including SVO, bio-diesel and bio-ethanol, is still in its infancy. Though there is considerable interest amongst various stakeholders on bio-fuels, the absence of a comprehensive policy on bio-fuels has impeded its progress. While SVO and bio-ethanol are commercially available, bio-diesel is available only in small quantities. At present, price of biofuels are not competitive compared with fossil fuels. But increase in the crude oil prices and technological advancements being made on the biofuel production can make these green fuels to economically vie with fossil fuels in the near future.

In order to diffuse bioethanol, it is necessary for government agencies to intervene in the related industries. Fuel standards and quality management systems are required to be arranged to make fuel quality more reliable, and in addition stable supply systems need to be built. Further, bio-ethanol must have price competitiveness against sugar and gasoline. Focused research for finding alternative for enhancing bio-ethanol production is needed. It can be said with a fair degree of certainty that till alternative feedstocks are made available, the future production and use of bio-ethanol for blending with petrol is likely to be rather limited. Apart from bio-ethanol production from cane sugar juice or its derivatives, bio-ethanol production based on lignocellulosic biomass will need to researched in a big way. Here again, the major issue is the production of bio-ethanol at a competitive price.

The present bio-diesel production comes from a few operational pilot plants and therefore, the market prices are high. The prices of bio-diesel will come down after a few years as harvest from the new plantations would become available. The projected prices of bio-diesel in various studies ranges from Rs. 16.50 per litre.

At present, there is no coordinated programme for development of jatropha plantation in the country. No reliable data on production and productivity of jatropha is available. The economic viability of bio-diesel will depend largely on the seed yields, which show large variations. In addition, due to absence of data from block plantations, several yield estimates are based on extrapolation of yields obtained from individual plants or small demonstration plots. The main challenge, as such, is to create an organized market for bio-diesel in country.

7.6.9 Hydrogen

In all likelihood, the energy shifts towards hydrogen is likely to be gradual. Hydrogen energy infrastructure would require a public policy framework for automotive and power generation purposes. The policy actions required may include development of codes and standards to support commercial activity; generate a critical mass in terms of resources, effort and competencies for "lighthouse" projects covering research, prototype development, technology demonstration, and commercialization leading to an application. In addition, creation of centres of excellence acting as focal points of critical research would be desirable.

7.7 Integration with National Development Goal

Renewable energy cannot be developed and disseminated in isolation. It has to fulfil national development obligations particularly in terms of access to energy services and contribution to overall energy security, by reducing demand of conventional fuels. Accordingly the targets and measures for renewables should be in with the overall developmental priorities.

7.8 Conclusion and the Way Forward

Renewable energy development is considered in India to be of great importance from the point of view of long-term energy supply security, environmental benefits and climate change mitigation. While assessing the contribution from renewable energy the Integrated Energy Policy Report has observed that contribution of modern renewables, excluding hydro would be around 5 to 6 per cent of India's energy mix by 2031-32. This appears to be not very encouraging a scenario and perhaps requires factoring in fast technological changes taking place in renewable energy. India, faced with twin challenges on energy and environmental fronts, has no option but to work towards increasing the role of renewables in the future energy systems. The challenges in hand are to increase the production volume for technology learning and for the cost reduction rate, which in turn affects the promotion policies. Larger application of renewable energy, then that of the present assessments, would depend on how rapidly the costs decline. Timely addressing of the most ambitious technical targets, *i.e.*, performance and cost reduction would an essential pre-condition to substantially increased renewable share in the total energy mix.

Continued and focused research and technology development would be essential and a mission mode approach to research would need to be planned for developing solar, bio-energy and hydrogen technologies. Investment in research would also be required to be increased manifold. India would also have to look for international cooperation in renewable energy through well defined R&D projects with proper division of labour and responsibilities for specific tasks with equitable financial burden and credit sharing arrangements. Bilateral, and to an extent multilateral scientific and technological cooperation agreements could provide framework for such R&D activities, including for an exchange of personnel and setting up of demonstration projects.

Technology plays a central role in addressing climate change issues. In this context India could mobilize global support for treating renewable energy technologies as 'global common' in the medium term. To begin with the these technologies could be placed in the public domain and joint research and development projects could be taken up between the institutions of developed and developing countries. The technology transfer cost could be fixed at no-profit level and the expenditure to be incurred in these acquisitions be made from global funds under climate change mechanisms. In parallel for addressing the intermittency of renewable power there is a need to work towards a global grid that could transmit electricity from one location to other from day to night locations and foster low-loss distribution.

The provisions of the Electricity Act, 2003 have already given a major boost for promotion of the renewable energy sector in India. All major States have announced feed-in-tariff for renewable power and the recent growth of renewable power is attributed to stable regulatory policy framework. However, there are many unresolved issues. As examined in Chapters 6 &7 implementation of the Indian renewable energy programme is still hampered by a number of market distortions and institutional, financial, and economic barriers. Right policy strategies and initiatives can help in overcoming these barriers and accelerating the current gradual transition to renewable energy. Further, renewable energy policy instruments would be required to have a complete synergy with the national climate policy.

Carbon finance and emissions trading are the longer-term incentives for developing renewable energy markets. These would require to be further strengthened. As a matter of fact renewable energy should automatically be considered to be "additional", and renewable energy developers should therefore expect to receive CERs for the fossil fuelled emissions that renewable energy displaces.

The Climate change has already projected that the renewables in medium and long term can be the option to make this planet livable. Although renewable energy has its own limitations and in a comparatively medium term these technologies could have a major role. The conclusion of the study is that renewable energy provides enormous benefits and can contribute significantly in the national energy mix at least economic, environmental and social costs. The time frame in which it can be achieved will depend on policies, commitment, and financial resources.

Index